INTRODUCTION TO CHEMICAL TRANSPORT IN THE ENVIRONMENT

Estimating the transport and fate of chemicals released into the environment is an interesting and challenging task. The global environment is large on the chemical transport and fate scale. This text applies the mathematics of diffusion, turbulent diffusion, and dispersion to the atmosphere, lakes, rivers, groundwater, and oceans, as well as transport between these media. The book follows a new educational paradigm of textbooks, in that it is based on examples and case studies. The required theory is explained as a technique for solving the case studies and example problems. A large portion of the book is dedicated to examples and case studies, from which the important principles are derived.

Dr. John S. Gulliver is the Joseph T. and Rose S. Ling Professor of Civil Engineering in the Department of Civil Engineering at the University of Minnesota, with an educational background in chemical engineering and civil engineering. His major engineering interests are in environmental fluid mechanics, chemical transport in environmental systems, and flow and chemical transport at hydraulic structures, on which he has published 98 peer-reviewed articles. He has investigated the measurement and prediction of air–water mass transfer at hydraulic structures, in river systems, at aerating hydroturbines, and in sparged systems and membrane aeration in reservoirs. He has investigated turbulent mixing and dispersion in lakes, reservoirs, and rivers and the fate and transport of a spilled nonaqueous phase liquid. He has developed numerical models to predict chemical and thermal transport and fate in rivers, reservoirs, and lakes. Dr. Gulliver has also advised on the efforts to reduce dissolved nitrogen concentrations downstream of dam spillways, consulted on techniques to remediate low dissolved oxygen concentrations that can occur in hydroelectric releases, and worked on forensic analysis of water quality problems that occur during operation of power facilities. He is co-editor of the *Hydropower Engineering Handbook*, *Air–Water Mass Transfer: Selected Papers from the Second International Symposium on Gas Transfer at Water Surfaces*, and *Energy and Sustainable Development Sub-Theme D, Proceedings of the 27th Congress of the International Association for Hydraulic Research*. He is currently the Coordinator of the Hydropower Institute. Dr. Gulliver received the Rickey Medal in 2003 from the American Society of Civil Engineers. He has been a visiting professor at the University of Karlsruhe, the University of São Paulo–São Carlos, Louisana State University, and the University of Chile, where he served as a Fulbright Scholar. He also served as a visiting research scientist at the Waterways Experiment Station of the U.S. Army Corps of Engineers.

Introduction to Chemical Transport in the Environment

JOHN S. GULLIVER

University of Minnesota

CAMBRIDGE UNIVERSITY PRESS
Cambridge, New York, Melbourne, Madrid, Cape Town, Singapore, São Paulo

Cambridge University Press
32 Avenue of the Americas, New York, NY 10013-2473, USA

www.cambridge.org
Information on this title: www.cambridge.org/9780521858502

First published 2007

Printed in the United States of America

A catalog record for this publication is available from the British Library.

Library of Congress Cataloging in Publication Data

Gulliver, John S.
Introduction to chemical transport in the environment / John S. Gulliver.
    p.   cm.
Includes bibliographical references and index.
ISBN-13: 978-0-521-85850-2 (hardback)
ISBN-10: 0-521-85850-X (hardback)
1. Environmental chemistry. 2. Diffusion. 3. Transport theory. 4. Pollution. I. Title.
TD193.G85 2006
628.5′2 – dc22   2006018209

ISBN 978-0-521-85850-2 hardback

Contents

Preface

This book is written as a text and reference for motivated seniors and first- or second-year graduate students in the area of chemical transport in the environment. The students in environmental sciences and engineering programs generally come from various backgrounds, such as chemical, civil, and mechanical engineering; chemistry; physics; biology; and environmental science. Courses are needed that focus on fundamentals of the environmental field, in this case environmental transport.

Emphasis is placed on developing the perspective and tools that will help students through graduate school and beyond. The diffusion equation is prevalent throughout environmental science and engineering, formulated to determine diffusion, turbulent diffusion, and dispersion of chemicals, in addition to convection. Learning about tools to solve this equation for different applications or boundary conditions is a task best undertaken early in one's career.

Without an environmental application, the material in this text may be perceived as dry and tedious. For that reason, much of the development of solution techniques is contained in examples of applications to environmental transport. Chapter 2, titled "The Diffusion Equation," is essential to the remainder of the text, which uses the examples provided in Chapter 2 to further develop solutions for other applications.

There are three reasons to develop analytical solutions to the diffusion equation during this age of numerical solutions. First, numerical solutions do not provide a good physical understanding of how diffusion works and what diffusion is. The insight gained through a serious study of the diffusion equation is difficult to achieve through another means. Second, it is handy to be able to perform back-of-the-envelope solutions to determine if a detailed numerical solution is justified. This can save a great deal of time. Third, numerical solutions are far from perfect. In fact, they are always wrong due to the need to discretize the solution domain into well-mixed cells in which the computer can add, subtract, multiply, and divide numbers to find an approximate solution to our boundary conditions. The question is not "what is the acceptable error" but "what is the error" in a numerical solution. The answer is best determined through developing an analytic solution to the diffusion equation with boundary conditions that are close to the application of interest and then solving the same problem computationally. The difference between the two is the true error in the numerical solution. At that point, a decision can be made about what error is acceptable.

This text is therefore designed to develop insight into the diffusion, turbulent diffusion, and dispersion problems that present themselves in chemical transport through the environment. It is also designed to provide skills that can be used to develop quick solutions to simplified boundary conditions for environmental transport problems. Finally, it is designed to test and verify computational solutions to the diffusion equation in situations that are similar to those to be modeled computationally.

1 The Global Perspective on Environmental Transport and Fate

Estimating the transport and fate of chemicals released into the environment is an interesting and challenging task. The environment can rarely be approximated as well mixed, and the chemicals in the environment often are not close to equilibrium. Thus, chemical transport and fate in the environment require a background in the physics of fluid flow and transport, chemical thermodynamics, chemical kinetics, and the biology that interacts with all of these processes. We will be following chemicals as they move, diffuse, and disperse through the environment. These chemicals will inevitably react to form other chemicals in a manner that approaches – but rarely achieves – a local equilibrium. Many times these reactions are biologically mediated, with a rate of reaction that more closely relates to an organism being hungry, or not hungry, than to the first- and second-order type of kinetics that we were taught in our chemistry courses.

To which environmental systems will these basic principles be applied? The global environment is large, on the chemical transport and fate scale. We will attempt to apply the mathematics of diffusion techniques that we learn to the atmosphere, lakes, rivers, groundwater, and oceans, depending on the system for which the material we are learning is most applicable. To a limited extent, we will also be applying our mathematics of diffusion techniques to transfer between these media. Volatilization of a compound from a water body, condensation of a compound from the air, and adsorption of a compound from a fluid onto a solid are all interfacial transport processes. Thus, the transport and fate of chemicals in the environmental media of earth, water, and atmosphere will be the topic. In this text, we will attempt to formulate transport and fate problems such that they can be solved, regardless of the media or the transport process, through the mathematics of diffusion.

A. Transport Processes

A transport process, as used herein, is one that moves chemicals and other properties of the fluid through the environment. *Diffusion* of chemicals is one transport process, which is always present. It is a spreading process, which cannot be reversed

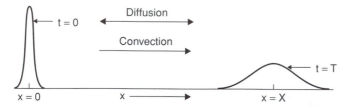

Figure 1.1. Illustration of convection and diffusion of a chemical cloud along the x-space coordinate (x-axis).

(without the involvement of another media such as in reverse osmosis). *Convection* or *advection* is the transport of chemicals from one place to another by fluid flow. The convection and diffusion of a chemical cloud, as represented in Figure 1.1, are the movements of the cloud and spreading of the cloud over time.

Turbulent diffusion is actually a form of advection, but the turbulent eddies tend to mix fluid with a random characteristic similar to that of the diffusion process, when viewed from enough distance. The representation given in Figure 1.1 could also be used to represent convection and turbulent diffusion, except that the pace of turbulent diffusion is normally more than one order of magnitude greater than diffusion. This higher pace of turbulent diffusion means that diffusion and turbulent diffusion do not normally need to be considered together, because they can be seen as parallel rate processes, and one has a much different time and distance scale from the other. If two parallel processes occur simultaneously, and one is much faster than the other, we normally can ignore the second process. This is discussed further in Section 1.D.

Dispersion is the combination of a nonuniform velocity profile and either diffusion or turbulent diffusion to spread the chemical longitudinally or laterally. Dispersion is something very different from either diffusion or turbulent diffusion, because the velocity profile must be nonuniform for dispersion to occur. The longitudinal dispersion of a pipe flow is illustrated in Figure 1.2. While there is diffusion of the chemical,

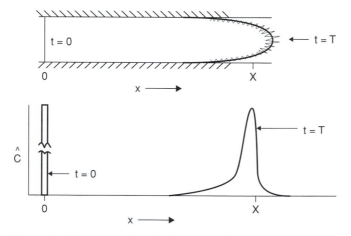

Figure 1.2. Illustration of longitudinal dispersion of a tracer "plane" at $t = 0$ to a dispersed "cloud" at $t = T$. $\hat{C}$ is the cross-sectional mean concentration.

the nonuniform velocity profile creates a dispersion that is much greater than would occur with diffusion alone. The other important difference is that *dispersion reflects the spreading of a cross-sectional mean concentration*, while diffusion represents the spreading of a local concentration. In some contexts, typically in atmospheric applications, turbulent diffusion is also considered to be a form of dispersion. This is only a semantic difference, and herein we will continue to distinguish between turbulent diffusion and the dispersion of a mean concentration.

Interfacial transfer is the transport of a chemical across an interface. The most studied form of interfacial transfer is absorption and volatilization, or condensation and evaporation, which is the transport of a chemical across the air–water interface. Another form of interfacial transfer would be adsorption and desorption, generally from water or air to the surface of a particle of soil, sediment, or dust. Illustration of both of these forms of interfacial transfer will be given in Section 1.D.

Finally, there is *multiphase transport*, which is the transport of more than one phase, usually partially mixed in some fashion. The settling of particles in water or air, the fall of drops, and the rise of bubbles in water are all examples of multiphase transport. Figure 1.3 illustrates three flow fields that represent multiphase transport.

Mass transport problems are solved with the *diffusion equation*, often represented as

$$\frac{\partial C}{\partial t} + u\frac{\partial C}{\partial x} + v\frac{\partial C}{\partial y} + w\frac{\partial C}{\partial z} = D\left[\frac{\partial^2 C}{\partial x^2} + \frac{\partial^2 C}{\partial y^2} + \frac{\partial^2 C}{\partial y^2} + \frac{\partial^2 C}{\partial z^2}\right] + S \quad (1.1)$$

1 | ← 2 → | | ← 3 → | 4

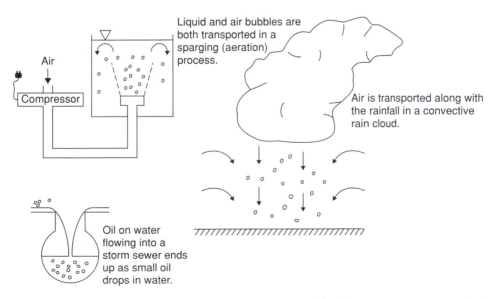

Figure 1.3. Illustration of multiphase transport. In these cases, air bubbles create a water flow and rain drops create an air flow. The oil drops do not have a significant rise or fall velocity in water and are simply transported.

where C is the concentration of a chemical; t is time; $u, v,$ and w represent the temporal mean velocity in the $x, y,$ and z directions, respectively; and D represents a diffusion coefficient. The first term (1) on the far left of equation (1.1) represents the rate of accumulation of chemical concentration. The second terms (2) represent the mean convection of the chemical. The third terms (3), to the right of the equal sign, represent either diffusion or turbulent diffusion of the chemical. The fourth term (4) represents the multitude of reactions that are possible in a fluid in environmental media.

We will be solving equation (1.1), or a similar equation, for various applications.

B. Chemical Fate

Chemical fate is the eventual short-term or long-term disposition of chemicals, usually to another chemical or storage. Some examples that fit the concept of short-term and long-term fate are given in Table 1.1. If a polychlorinated biphenol (PCB) compound is in groundwater, the media are soil and water. The short-term fate will be that the PCB will primarily adsorb to the soil. The long-term fate is that the chemical will desorb, when the PCB-laden water has left, and eventually be bioremediated by microbacteria looking for carbon sources. If this PCB is in the atmosphere, it will be adsorbed primarily to aerosols and particles in the short term, whereas its long-term fate will probably be photocatalyzed degradation.

There are as many or more examples of short-term and long-term fate as there are chemical–media combinations. An important consideration for this topic is whether we are interested in short-term or long-term fate. This is often a question to be answered by toxicologists. We will, for example, take the results of their computations and experiments and track the more toxic forms of a spill. Sometimes this involves a short-term fate, and sometimes this involves a long-term fate. The time scale of the calculations is important in determining how we deal with the problem or how we set up our solution.

Table 1.1: *Examples of short-term and long-term fates*

Chemical	Media	Short-term fate	Long-term fate
PCB	Soil and water	Adsorbed to soil	Biomediated degradation
PCB	Atmosphere	Adsorbed to aerosols	Photocatalyzed degradation
CO_2	Water	Reactions to carbonate and bicarbonate	Photosynthesis to oxygen and biomass
Benzene	Water	Adsorbed to suspended particles	Bioremediated degradation
Ammonia	Soil and water	Reaction to ammonium	Bioremediated degradation to nitrogen

PCB, polychlorinated biphenyl.

C. The Importance of Mixing

Mixing is a rate-related parameter, in that most rates of reaction or transport are dependent on mixing in environmental systems. When mixing is dominant (the slowest process), the first-order rate equation can be described as

$$\text{Rate of process} = \text{Mixing parameter} \times \text{Difference from equilibrium} \qquad (1.2)$$

Thus, we need two items to compute the rate of the process: the equilibrium concentrations for all species involved and the mixing rate parameter. A common example would be dissolved oxygen concentration in aquatic ecosystems.

One of the most common chemicals of concern in water bodies is oxygen. Without sufficient oxygen, the biota would be changed because many of the "desirable" organisms in the water body require oxygen to live. The rate of oxygen transfer between the atmosphere and a water body is therefore important to the health of the aquatic biota. For air–water oxygen transfer, equation (1.2) can be formulated as

$$\frac{dM}{dt} = K_L \, A \left(\frac{C_a}{H} - C \right) \qquad (1.3)$$

where dM/dt is the rate of mass transfer into the water, K_L is a bulk oxygen transfer coefficient, A is the surface area for transfer, C_a is the concentration of oxygen in the air, H is a coefficient that partitions oxygen between the air and water at equilibrium (called Henry's law constant for liquid and gas equilibrium), and C is the concentration of oxygen in the water. Air is approximately 20.8% oxygen, so the concentration of oxygen in the atmosphere is determined primarily by atmospheric pressure. Henry's law constant for oxygen is a function of pressure as well as temperature. Thus, the equilibrium concentration of oxygen is influenced by the thermodynamic variables: pressure and temperature. The rate parameter is $K_L A$, which has units of volume per second. The difference from equilibrium partitioning is represented by $C_a/H - C$. It is C that we typically need to bring as close to equilibrium with the atmosphere as possible, and the means to do it is by having a large dM/dt. This usually means a large $K_L A$ because it would be difficult to alter either C_a or H in the atmosphere. While the surface area is often established by the boundary conditions, K_L is determined by turbulence and diffusion coefficient (i.e., mixing) close to the water surface and represents the rate of mixing per unit surface area. Thus, the primary variable that can be changed in order to increase dM/dt is the mixing parameter represented by K_L. Some further examples of mixing rate and equilibrium parameters in environmental processes are given in Table 1.2.

D. Resistance to Transport

An important concept for environmental transport is *resistances*. The inverse of a rate parameter is a resistance to chemical transport. Or, in equation form:

$$1/\text{Rate parameter} = \text{Resistance to chemical transport} = R \qquad (1.4)$$

Table 1.2: *Examples of important mixing rate and equilibrium parameters in environmental process*

Process	Mixing rate-related parameter	Equilibrium parameter
1. Treatment Processes		
Coagulation/flocculation	Size of coagulation and flocculation basins and proper mixing (residence time)	Dose of coagulants (alum)
Softening	Design of softening tank to increase mixing	Dose of softening agent (lime)
Settling	Design of settling basin to reduce mixing	$\rightarrow 0$
Chlorination	Design of chlorination and dechlorination chambers for proper mixing and residence time	Dose of chlorine
Filtration	Size of filter bed	Length of time before backflushing
2. Surface Waters		
Oxygen transfer	Diffusion and turbulent mixing	Atmospheric concentration of oxygen and Henry's law constant
Volatilization of pollutants	Diffusion and turbulent mixing	$\rightarrow 0$
Toxic spills	Diffusion and turbulent mixing	Spill-water equilibrium
Internal loading of nutrients	Hypolimnetic mixing	Oxygen concentration in hypolimnion
Sorption onto suspended sediments	Turbulent mixing exposes chemicals to sediment	Sediment–water partitioning
3. Atmosphere		
Greenhouse gases (CO_2, CH_4)	Turbulent mixing	Atmospheric concentrations
Volatilization of spills	Turbulent mixing–dispersion	$\rightarrow 0$
Aerosols	Turbulent mixing – dispersion – settling	None
4. Groundwater and Sediments		
Spills	Advection – dispersion	$\rightarrow 0$
Oxygen	Diffusion – advection	Atmospheric concentration of oxygen
Sorption of soil	Advection – diffusion	Equilibrium soil–water partitioning

Figure 1.4 gives an example of the adsorption of a compound to suspended sediment, modeled as two resistances in series. At first, the compound is dissolved in water. For successful adsorption, the compound must be transported to the sorption sites on the surface of the sediment. The inverse of this transport rate can also be considered as a resistance to transport, R_1. Then, the compound, upon reaching the surface of the suspended sediment, must find a sorption site. This second rate parameter is more related to surface chemistry than to diffusive transport and is considered a second resistance, R_2, that acts in series to the first resistance. The second resistance cannot

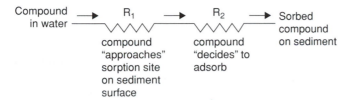

Figure 1.4. Adsorption analogy to two resistors in a series: adsorption of an organic compound to sediment.

occur without crossing the first resistance of transport to the sorption site; so, they must occur in series.

Now, if R_1 is much greater than R_2, we can assume that R_2 is zero without compromising the accuracy of the rate calculation. In electric circuits, two resistances applied in series are simply added together in calculating the line resistance. The same is true for resistance to chemical transport. If R_1 is 1,000 resistance units and R_2 is 1 resistance unit, we can ignore R_2 and still be within 99.9% of the correct answer. For most environmental transport and fate computations, it is sufficient to be within 99.9% of the correct answer.

Another example is the air–water transfer of a compound, illustrated in Figure 1.5. This example will be used to explain volatile and nonvolatile compounds. There is resistance to transport on both sides of the interface, regardless of whether the compound is classified as volatile or nonvolatile. The resistance to transport in the liquid phase is given as $R_L = 1/K_L$. If we are describing chemical transfer through an equation like (1.3), the resistance to transfer in the gas phase is given as $R_G = 1/(HK_G)$. The equilibrium constant is in the R_G equation because we are using the equivalent water side concentrations to represent the concentration difference from

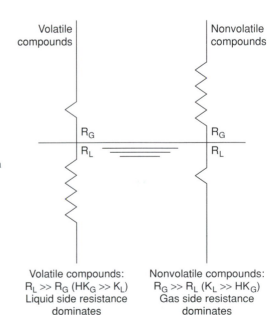

Figure 1.5. Air–water transfer analogy to two resistors in a series.

equilibrium, and the gas phase resistance needs to be a resistance to an equivalent water concentration.

The gas phase and the liquid phase resistances are applied in series. In general, gas film coefficients are roughly two orders of magnitude greater than liquid film coefficients. It is also true that Henry's law constant, H, varies over many orders of magnitude as the transported compounds are varied. Nitrogen gas, for example, has a Henry's law constant of approximately 15, using mass concentrations. The herbicide atrazine has a Henry's law constant of 3×10^{-6}. Thus, the ratio R_G/R_L would vary by seven orders of magnitude between nitrogen gas and atrazine (see Appendix A–5).

If we put these orders of magnitude into a series resistance equation

$$R = R_L + R_G = \frac{1}{K_L} + \frac{1}{HK_G} \tag{1.5}$$

Because of Henry's law constants, we can see that, for nitrogen gas, $R \cong R_L$, and for atrazine, $R \cong R_G$. If the ratio of $K_G/K_L \sim 100$ is applied (Mackay and Yuen, 1983), $R_G = R_L$ when $H = 0.01$.

Now, the mass transfer between phases is given as

$$\frac{dM}{dt} = \frac{A}{R}\left(\frac{C_a}{H} - C\right) \tag{1.6}$$

or

$$\frac{dM}{dt} = \frac{A}{\dfrac{1}{K_L} + \dfrac{1}{HK_G}}\left(\frac{C_a}{H} - C\right) \tag{1.7}$$

Nitrogen gas would be a volatile compound, because the equilibrium is strongly to the gas phase, and there is little gas phase resistance to its transfer (that is, $1/K_L >> 1/(HK_G)$). For that reason, nitrogen is generally called a gas, as are many other volatile compounds, such as methane, oxygen, and propane.

Atrazine, however, would be a nonvolatile compound – $1/(HK_G) >> 1/K_L$ – because equilibrium is strongly to the liquid phase due to the small Henry's law constant. There is also a strong gas phase resistance to the transfer. Atrazine was manufactured to remain in the liquid phase, where it will act as a herbicide, rather than in the gas phase, where farm personnel will be breathing this toxic chemical. If you were going to pick a compound that is *not* made by humans from the list of those that are a gas or liquid in our environment, a good guess is that it would be a volatile or semivolatile compound. There are only a few nonionic environmental compounds that are nonvolatile. Remarkably, one of them is water. While the atmosphere may be as much as 3% water, the water bodies in the world are very close to 100% water. The equilibrium is strongly to the liquid side because of the small Henry's law constant.

One theme of this discussion can now be stated as follows: when transport processes occur in series, it is the slower transport processes that are important for chemical transport calculations, because the *resistance* to transport is large, just as the large resistors of a series in an electronic circuit are the most important.

Now we are ready for the second theme: when transport processes *occur in parallel*, the fast transport process with the low resistance dominates. The result is the

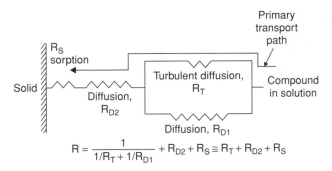

Figure 1.6. Transport to a sorption site and the resistor analogy.

opposite of resistances in series. Figure 1.6 illustrates this concept with the transport of a compound from the water body to a sorption site on a solid. In the bulk solution, there is diffusion and turbulent diffusion occurring simultaneously. Transport can occur from either process, so there are two different paths that may be followed, without the need of the other path. These transport processes are operating in parallel, and the faster transport process will transport most of the compound. The analogy to electronic circuits applies in this case as well. Beginning with a compound in solution in Figure 1.6, there are two parallel transport paths, each with a resistance to transfer. Most of the compound will be transported through the path with the least resistance. Many times, we can ignore the path with the greater resistance because the quantity of compound transported through this path is very small. When the compound comes close to the solid, however, the turbulent diffusion dissipates, because eddies become so small that they are dissipated by viscous action of the water. Now, we are back to one transport path, with the act of sorption and diffusion acting in series. Thus, the slowest transport path once again becomes the important process.

The overall resistance to the sorption process illustrated in Figure 1.6 can be written as follows:

$$R = \frac{1}{1/R_T + 1/R_{D1}} + R_{D2} + R_S \cong R_T + R_{D2} + R_S \qquad (1.8)$$

where $R_T, R_{D1}, R_{D2},$ and R_S are the resistance to turbulent transport, diffusive transport in the bulk of the fluid, diffusive transport near the solid surface, and adsorption, respectively. We can see that, in Figure 1.6 and in equation (1.8), the resistance from diffusion in the bulk of the fluid can be neglected because turbulent diffusion is a parallel path. The resistance from diffusion only needs to be considered when there is no parallel path for turbulent diffusion, such as very near the surface of the solid. Thus, we can ignore R_{D1}, but not R_{D2}.

In this chapter, we have discussed some of the topics in the bulk of the text, where the physics of mass transport – rather than the mathematics of the diffusion equation – are essential. We will return to these and similar engineering concepts throughout the text in an attempt to develop models in the environmental transport and fate of chemicals that are realistic but can be solved, even if that solution is approximate.

E. Terminology of Chemical Transport

There is some terminology of chemical transport that needs to be defined. Some terminology has already been implied, and some may seem obvious, but these are so important to this text that we need to define them in one place and at one time.

Diffusion – In this text, we will define diffusion (and most other processes) from an engineering perspective, in that we will go to the level of detail that suits our objective. Diffusion can then be defined as the mixing of chemicals by random molecular motion. Diffusion coefficients in dilute solutions will be discussed in detail in Chapter 3.

Convection (or advection) – The transport of a chemical or other quantity by an imposed flow.

Turbulent diffusion – The mixing of chemicals by turbulence, such that a turbulent diffusion coefficient can be defined separately from the temporal mean convection.

Dilution – The mixing of a more concentrated solution with one that is less concentrated. The adage "The solution to pollution is dilution" is still used, sometimes appropriately, for many pollution and mitigation processes.

Density – Total mass per unit volume.

Concentration – The quantity of a compound or chemical per unit volume, unit mass, or unit moles, where 1 mole $= 6.02 \times 10^{23}$ molecules of the chemical or compound. In this text, we will typically be discussing concentration in mass or moles per volume of water, mass per mass of solid, and moles per mole of gas, depending on the media of interest.

The conversion between concentration units and the expression of the units themselves can be confusing. We will now review the typical concentration units used in various environmental media. *Concentration in water* is usually given as mass per unit volume or moles per unit volume. The conversion between them is a straightforward application of molecular weights. For example, we have 2.0 g/m^3 of CO_2 dissolved in water. The molecular weight of carbon dioxide is 44 g/mole. Then the concentration in moles/m^3 is

$$\frac{2 \text{ g/m}^3 \text{ CO}_2}{44 \text{ g/mole}} = 0.0455 \text{ moles/m}^3 \text{ of CO}_2 \qquad (1.9)$$

The *concentration in air*, however, is typically given in units that are different from those of water, because mass per unit volume can be misleading in a media that can be significantly compressed. Thus, concentration in the atmosphere is often given as a partial pressure at one atmosphere of total pressure. Because the pressure of a gas at a given temperature is proportional to the number of molecules in a given volume, the following relations are applied:

$$\frac{\text{Partial pressure}}{\text{Atm. pressure}} = \frac{\text{Molecules of compound}}{\text{Total molecules}} = \frac{\text{Moles of compound}}{\text{Total moles}} \qquad (1.10)$$

Table 1.3: *Common abbreviations for concentration units*

Abbreviation	Definition	Units	In water
ppm	parts per million by weight	mg/kg	mg/L or g/m^3
ppb	parts per billion by weight	μg/kg	μg/L or mg/m^3
ppt	parts per trillion by weight	ng/kg	ng/L or μg/m^3
			For gases
ppm(v)	parts per million by volume	mL/m^3	10^{-6} atm/atm
ppb(v)	parts per billion by volume	μg/m^3	10^{-9} atm/atm
ppt(v)	parts per trillion by volume	ng/m^3	10^{-12} atm/atm

Finally, *concentration units in soil* are a combination of the above with a twist. Soil is a multiphase media, and different concentration units are applied, typically, to each phase. The interstitial space (in between the soil) is filled with water or air in most cases, and the corresponding concentration units for water or air, respectively, would be used. In addition, the concentration of a compound adsorbed to the sediment is normally given either as mass of compound/mass of solid or moles of compound/mass of solid. The mass of solid is one of the few soil parameters that can be determined definitively, so that is what is used.

There are some abbreviations for concentration units that are often seen in the literature and will be used periodically in this text. These are listed in Table 1.3. Also listed is a common conversion to water or to air.

EXAMPLE 1.1: *Partial pressure determination*

There is some air that is 2 ppm(v) methane. Determine the partial pressure of methane at an atmospheric pressure of 1 atm.

For gases, at pressures that are not too far from atmospheric, the space occupied by one molecule of methane is equal to that occupied by one molecule of nitrogen or oxygen gas. We can therefore use the ideal gas law to convert ppm(v) to atmospheres of methane/atmosphere of total pressure.

$$P_{air} V = n_{air} RT \tag{E1.1.1}$$
$$P_{meth} V = n_{meth} RT \tag{E1.1.2}$$

where P is pressure or partial pressure, V is volume, n is the moles of a compound contained in the volume V, R is the universal gas constant, and T is absolute temperature. Dividing equation (E1.1.2) by equation (E1.1.1) results in the ratio:

$$\frac{P_{meth}}{P_{air}} = \frac{n_{meth}}{n_{air}} \tag{E1.1.3}$$

Because the volume occupied by each molecule is similar at low pressures, the concentration by volume is equal to the concentration by moles, and $n_{meth}/n_{air} = 2 \times 10^{-6}$. Therefore, $P_{meth}/P_{air} = 2 \times 10^{-6}$. This means that the two are interchangeable. At one atmosphere total pressure, then, the methane partial pressure for this example is 2×10^{-6} atm. Whether we are talking about partial pressures or concentration by volume, they will have the same value in a gas as long as the ideal gas law is applicable.

F. Definition of Means

Mean values are important in environmental transport and fate because the environment is not well mixed. To address various applications in the most effective manner, we often consider mean values and the variations from the mean values separately. We will be predominantly using two types of mean values: temporal means and cross-sectional means.

A *temporal mean* is the mean of a fluctuating quantity over a time period, T. If the time period is sufficiently long, the temporal mean values are constant over time. Temporal means are often used in analyzing turbulent diffusion. For example, if u is the x-component of velocity and is a function of space and time, $u = u(x, y, z, t)$, in cartesian coordinates. Then the temporal mean velocity, $\bar{u}$, would only be a function of x, y, and z:

$$\bar{u}(x,y,z) = \frac{1}{T} \int_{t=0}^{T} u(x,y,z,t)dt \qquad (1.11)$$

The concentration, under these conditions, would likely be a function of time as well, such that the temporal mean concentration, $\bar{C}$, would also be required:

$$\bar{C}(x,y,z) = \frac{1}{T} \int_{t=0}^{T} C(x,y,z,t)dt \qquad (1.12)$$

where x, y, and z are the spatial coordinates and t is time. Taking the temporal mean values of equations (1.11) and (1.12) greatly simplifies the solution of the diffusion equation in turbulent flow.

A *cross-sectional mean* is the mean value of a quantity over a cross section. We will use the cross-sectional means to compute concentration in a system with dispersion. An illustrative example is given in Figure 1.7. For the system visualized in this figure, the cross-sectional mean velocity, $U(x, t)$, and cross-sectional mean concentration, $\hat{C}(x, t)$, would be given by

$$U(x, t) = \frac{1}{A_{CS}} \int_{y} \int_{z} u(x, y, z, t)dy\, dz \qquad (1.13)$$

and

$$\hat{C}(x, t) = \frac{1}{A_{CS}} \int_{y} \int_{z} C(x, y, z, t)dy\, dz \qquad (1.14)$$

Random fluctuations in C and u are normally smoothed out in time as well, because these cross-sectional mean values also act similar to a low-pass filter in time. U and $\hat{C}$ can change slowly, but not rapidly, because of the general smoothing character of

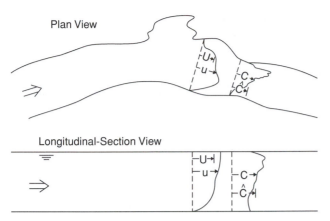

Figure 1.7. Illustration of cross-sectional mean velocity, U, and cross-sectional mean concentration, $\hat{C}$, versus u and C.

cross-sectional averages with regard to temporal variations that appear random in time and space.

G. Applications of Topical Coverage

This text is organized by physical transport and fate topics, not by application to a given system. For example, solutions to the diffusion equation with different boundary conditions is the topic of Chapter 2. These solutions are generally applied to sediment, groundwater, and biofilms, as well as to other systems. We will be discussing these applications throughout the text, but it may be difficult to find a section that deals solely with groundwater. The solution techniques are quite similar, and one objective is to emphasize that these solutions have many different applications. A solution to a chemical transport problem in biofilms may also be applied to sediments, and often, with some minor adjustments, to groundwater flow. But, in order to bring some coherence to the text, the following description provides the systems to which the topics in the text are most often applied.

Chapter 2: The Diffusion Equation. The diffusion equation provides the mathematical foundation for chemical transport and fate. There are analytical solutions to the diffusion equation that have been developed over the years that we will use to our advantage. The applications in this chapter are to groundwater, sediment, and biofilm transport and fate of chemicals. This chapter, however, is very important to the remainder of the applications in the text, because the foundation for solving the diffusion equation in environmental systems will be built.

Chapter 3: Diffusion Coefficients. This chapter demonstrates how to estimate the diffusion coefficients of dilute chemical concentrations in water and air. The chapter is important any time that diffusion cannot be ignored in an application of chemical transport and fate. Some of these cases would be in laminar flows, in sediment, in groundwater transport, and close to an interface in turbulent flows.

Chapter 4: Mass, Heat, and Momentum Transport Analogies. The transport of mass, heat, and momentum is modeled with analogous transport equations, except for the source and sink terms. Another difference between these equations is the magnitude of the diffusive transport coefficients. The similarities and differences between the transport of mass, heat, and momentum and the solution of the transport equations will be investigated in this chapter.

Chapter 5: Turbulent Diffusion. Turbulent diffusion is an important transport mechanism in the atmosphere, oceans, lakes, estuaries, and rivers. In fact, most of the atmosphere and surface waters of the Earth are turbulent. If you are going to work in any of these systems, it will be important to have at least a working knowledge of turbulent diffusion.

Chapter 6: Reactor Mixing Assumptions. Neither a perfectly mixed system nor a system without mixing exists, but these reactor mixing assumptions can often be used to help us get to an approximate solution to our diffusion problem. The most obvious application of reactor mixing is to treatment systems for water and wastewater. All of the concepts of reactor mixing, however, are also applied to the environment for many transport and fate problems. A well-mixed lake is often assumed, for example, for the mass balance of some chemicals, and a plug flow river has also been assumed for many applications since Streeter and Phelps originally developed the first solution for the oxygen sag equations. The assumptions of many reactor models eliminate the diffusive term in the diffusion equation, greatly simplifying analytical and numerical solutions. For complex mass balances this simplification is convenient.

Dispersion is another reactor mixing topic that will be discussed in Chapter 6. Dispersion normally is used when cross-sectional mean concentrations and velocities are being computed. A cross-sectional mean concentration is useful for a pipe, stripping tower, river, or groundwater transport.

Chapter 7: Computational Mass Transport. Computational mass transport is a more flexible solution technique that still requires verification with an analytical solution. A short description of some of the more prevalent techniques applied to mass transport is provided in this chapter.

Chapter 8: Interfacial Mass Transfer. Transfer of mass between two phases – such as air–water, water–solid, and air–solid – is considered in this chapter. The emphasis is on the development of theories to describe the concentration boundary layer for various applications.

Chapter 9: Air–Water Mass Transfer in the Field. The theory of interfacial mass transfer is often difficult to apply in the field, but it provides a basis for some important aspects of empirical equations designed to predict interfacial transport. The application of both air–water mass transfer theory and empirical characterizations to field situations in the environment will be addressed.

H. Problems

1. Human exposure limits through air exposure are to be applied at a plant that manufactures Heptachlor, $C_{10}H_5Cl_7$. Heptachlor is a chemical that is currently restricted in use to termite control, because of exposure concerns. These limits are given in ppm(v), but the release to the air is determined in g/s. Applying the volume flow with the current ventilation system and the volume of the plant in a complete mix reactor assumption resulted in a vapor phase concentration of 0.1 $\mu g/m^3$. What is the vapor phase concentration in ppm(v)?

2. Describe the difference between diffusion, turbulent diffusion, and dispersion.

3. Apply the resistance to transport analogy to describe sediment–water column transfer of a nonreactive compound.

4. Describe the difference between the mathematics of temporal and cross-sectional mean values of velocity and concentration. How does the cross-sectional mean also operate similar to a low-pass filter (removes high-frequency fluctuations) in time?

2 The Diffusion Equation

In this chapter, we will review various solution techniques for the diffusion equation, which is generally defined as the mass transport equation with diffusive terms. These techniques will be applied to chemical transport solutions in sediments. There are also a number of applications to chemical transport in biofilms. There are many other applications of the diffusion equation, including most of the topics of this text, but they require more background with regard to the physics of mixing processes, which will be addressed in later chapters.

What is *mass* (or *chemical*) *transport*? It is the transport of a solute (the dissolved chemical) in a solvent (everything else). The solute is the dissolvee and the solvent is the dissolver. There are liquids that are generally classified as solvents because they typically play that role in industry. Some examples would be degreasing and dry cleaning solvents, such as trichloroethylene. In environmental applications, these "solvents" are the solutes, and water or air is usually the solvent. In fact, when neither water nor air is the solvent, the general term *nonaqueous phase liquid* is applied. A nonaqueous phase liquid is defined as a liquid that is not water, which could be composed of any number of compounds.

The substance being transported can be either dissolved (part of the same phase as the water) or particulate substances. We will develop the diffusion equation by considering mass conservation in a fixed control volume. The mass conservation equation can be written as

$$\underset{\text{IN}}{\text{Flux rate}} - \underset{\text{OUT}}{\text{Flux rate}} + \underset{\text{rate}}{\text{Source}} - \underset{\text{rate}}{\text{Sink}} = \text{Accumulation} \tag{2.1}$$

Now that we have our mass conservation equation, we must decide which control volume would be the most convenient for our applications. The control volumes used most for this type of mass balance are given in Figure 2.1. The general control volume, given in Figure 2.1a, is used for descriptive purposes, to maintain generality. It is rare that we actually work with something that approximates such a contorted control volume. The control volumes that are used in practice are given in Figure 2.1b–d. For the environmental applications of chemical transport, the rectangular control

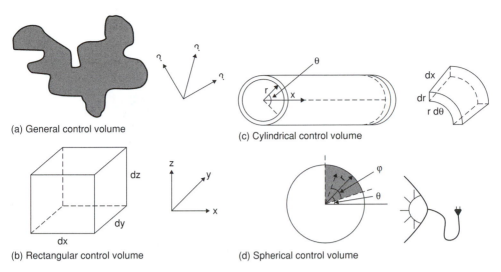

(a) General control volume

(c) Cylindrical control volume

(b) Rectangular control volume

(d) Spherical control volume

Figure 2.1. Common control volumes found in engineering texts and (for the latter three) used in solving the diffusion equation.

volume (Figure 2.1b) has proven to be the most useful. The cylindrical control volume (Figure 2.1c) is used to make pipe or tube flow problems easier to solve, and the spherical control volume (Figure 2.1d) is often helpful when dealing with transport in and around particles or drops. For this control volume, it is convenient to imagine a light being shined along the axis, which casts a shadow of the vector onto a plane normal to the light. The φ angle measures from the reference axis to the shadow in this plane.

We will use the rectangular control volume for the development of our mass conservation (diffusion) equation.

A. Development of the Diffusion Equation

The diffusion equation will be developed by considering each term in equation (2.1) separately. In addition, the flux terms will be divided into diffusive and convective flux rates.

1. Diffusive Flux Rate

The molecules of a fluid "at rest" are still moving because of their internal energy. They are vibrating. In a solid, the molecules are held in a lattice. In a gas or liquid, they are not, so they move around because of this vibration. Since the molecules are vibrating in all directions, the movement appears to be random.

Let us look at one face of our rectangular control volume (our box) and imagine that we put a tracer on the outside of the box, as shown in Figure 2.2. Initially, the tracer molecules will be distributed uniformly on the outside of the box, with a concentration distribution as shown in Figure 2.2a. However, all of these molecules are vibrating, with inertial movements back and forth. If we look at the tracer

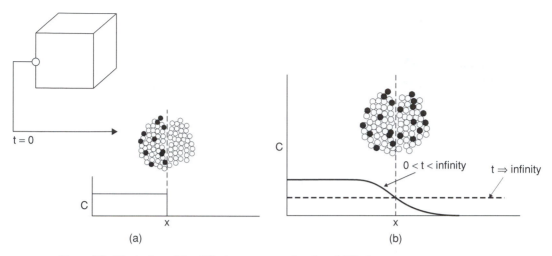

Figure 2.2. Illustration of the diffusion process and scales of diffusion.

molecules after a certain time, this apparently random motion would have distributed them throughout the box, such that there is no concentration gradient inside or outside the box. This is illustrated in Figure 2.2b.

Then, why is diffusion such a slow process if the molecules are all vibrating and exchanging places? Let us back up and observe the scale of diffusion for the example of liquid water as the solvent. One mole of water is equal to 6.02×10^{23} molecules (Avogadro's number) and will weigh 18 g. Since 1 g of water occupies roughly 1 cc, we have 18 cc's of water, which can be represented by a cube that is 2.6 cm on a side. The cube root of 6×10^{23} is approximately 8×10^7. Thus, we can visualize one corner of our 2.6 cm cube as being occupied by 80 million molecules. It takes some time for the random motion of these molecules, regardless of how active, to traverse 80 million molecules by a motion that appears random. Diffusion is especially slow in liquids because the molecules are small, compared with the distance traversed, and are relatively close together.

The molecules are generally much farther apart in gases, so the diffusivity of a compound in a gas is significantly larger than in a liquid. We will return to this comparison of diffusion in gases and liquids in Chapter 3.

Fick's Law. Fick's law is a physically meaningful mathematical description of diffusion that is based on the analogy to heat conduction (Fick, 1855). Let us consider one side of our control volume, normal to the x-axis, with an area A_x, shown in Figure 2.3. Fick's law describes the diffusive flux rate as

$$\text{Diffusive flux rate} \quad = \quad -D \qquad \frac{\partial C}{\partial x} \qquad A_x$$

$$\text{(g/s)} \qquad \text{(m}^2\text{/s)} \qquad \text{(g/m}^4\text{)} \qquad \text{(m)}^2$$

$$(2.2)$$

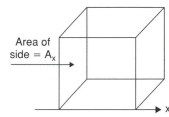

Figure 2.3. Illustration of net diffusive flux through one side of the rectangular control volume.

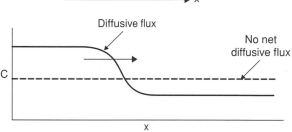

where C is concentration of the solute (tracer); D is the diffusion coefficient of the solute in the solvent (water), which relates to how fast how and far the tracer molecules are moving to and fro; and $\partial C/\partial x$ is the gradient of concentration with respect to x, or the slope of C with x, as shown in Figure 2.3. Thus, the diffusive flux rate depends on the diffusion coefficient and the *gradient* of concentration with distance. In Figure 2.2a there was a greater gradient of tracer molecules than in Figure 2.2b, so there would be a larger chemical flux across the surface of the control volume. The same molecular motion would bring more tracer molecules into the box in Figure 2.2a than in Figure 2.2b, especially when we realize that the diffusive flux is a net flux. Any molecules that come back out of the box, after entering, would count against the diffusive flux into the box.

2. Convective Flux

The convective flux rate into our control volume is simply the chemical mass carried in by convection. If we consider the same box of Figure 2.3, except with a velocity component u in the x-direction, the convective flux rate into the box from the left-hand side is

Convective flux rate	=	Velocity component normal to surface	×	Surface area	×	Concentration	(2.3)
(g/s)		(m/s)		(m²)		(g/m³)	

or

$$\text{Convective flux rate} = u A_x C \qquad (2.4)$$

where u is the component of velocity in the x-direction and A_x is the surface area normal to the x-axis on that side of the box. All six sides of our box would have a convective flux rate through them, just as they would have a diffusive flux.

3. Rate of Accumulation

The rate of accumulation is the change of chemical mass per unit time, or

$$\text{Rate of accumulation} \quad = \quad \overline{V} \quad \frac{\partial C}{\partial t} \qquad (2.5)$$

$$\text{(g/s)} \qquad\qquad \text{(m}^3\text{)} \quad \text{(g/m}^3\text{/s)}$$

where $\overline{V}$ is the volume of our box.

4. Source and Sink Rates

The solute chemical can appear or disappear through chemical reaction. In addition, interfacial transfer is often integrated over the control volume and considered as a source or sink throughout the control volume (see Example 5.4). This type of pseudo-reaction can be of significant help in solving chemical transport problems when averages over a larger control volume, such as cross-sectional mean concentrations, are being computed. For both cases (chemical reactions and pseudo-reactions), the source and sink rates are given as

$$\text{Source} - \text{sink rate} = S \qquad \overline{V} \qquad\qquad (2.6)$$

$$\text{(g/s)} \qquad \text{(g/m}^3\text{/s)} \qquad \text{(m}^3\text{)}$$

where S is the net source/sink rate per unit volume. The particular reactions that a given chemical is likely to undergo will determine the form of S used in equation (2.6). These are listed in Table 2.1. The source/sink term could be a combination of two or more of these reactions. For convenience in determining analytical solutions to the diffusion equation, most source/sink terms are approximated as either a first-order or zero-order reaction. This type of application will be demonstrated in Section 2.F.

Table 2.1: *Common source and sink terms used in the diffusion equation*

Source/sink name	Equation	Units of constant
Zero order	$S = k_0$	$k_0 = \text{g/m}^3 - \text{s}$
First order	$S = k_1 C$	$k_1 = \text{S}^{-1}$
Second order	$S = k_2 C^2$	$k_2 = \text{m}^3/\text{g} - \text{s}$
Independent variable	$S = k_{1i} P^*$	$k_{1i} = \text{s}^{-1}$
	$S = k_{2i} PC^\dagger$	$k_{2i} = \text{m}^3/\text{g} - \text{s}$
Monod kinetics‡	$S = \frac{\mu_m C}{k_c + C} P$	$\mu_m = \text{maximum growth rate (s}^{-1}\text{)}$ $k_c = \text{half-saturation coefficient (g/m}^3\text{)}$

* If P is nearly constant, then k_{1i} can be provided as a zero-order term.
† Often called second order.
‡ Common for biologically mediated reactions.

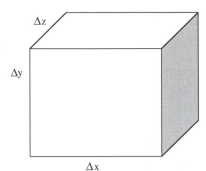

Figure 2.4. Dimension of the rectangular control volume.

5. Mass Balance on Control Volume

A mass balance on one compound in our box is based on the principle that whatever comes in must do one of three things: (1) be accumulated in the box, (2) flux out of another side, or (3) react in the source/sink terms. If it seems simple, it is.

　　We will begin by assigning lengths to the sides of our box of dx, dy, and dz, as shown in Figure 2.4. Then, for simplicity in this mass balance, we will arbitrarily designate the flux as positive in the $+x$-direction, $+y$-direction, and $+z$-direction. The x-direction flux, so designated, is illustrated in Figure 2.5. Then, the two flux terms in equation (2.1) become

$$\text{Flux rate in} + \text{Difference in flux rate} = \text{Flux rate out} \qquad (2.7)$$

or, because a *difference* can be equated to a *gradient times the distance over which the gradient is applied,*

$$\text{Flux rate out} - \text{Flux rate in} = \text{Gradient in flux rate} \times \text{Distance} \qquad (2.8)$$

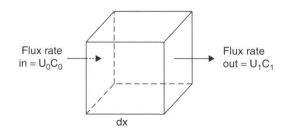

Figure 2.5. Illustration of the x-component of mass flux rate into and out of the rectangular control volume.

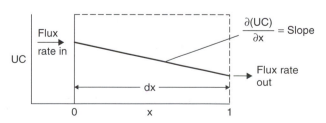

Equation (2.8) can thus be applied along each spatial component as

$$\text{Flux rate (out − in)}_x = \frac{\partial}{\partial x}(\text{flux rate})dx \qquad (2.9a)$$

$$\text{Flux rate (out − in)}_y = \frac{\partial}{\partial y}(\text{flux rate})dy \qquad (2.9b)$$

$$\text{Flux rate (out − in)}_z = \frac{\partial}{\partial z}(\text{flux rate})dz \qquad (2.9c)$$

Convective Flux Rates. We will deal with the convective and diffusive flux rates separately. They will eventually be separated in the final diffusion equation, and it is convenient to make that break now. The x-component of the convective flux rate is equal to the x-component of velocity, u, times the concentration, C, times the area of our box normal to the x-axis. Therefore, in terms of convective flux rates, equation (2.9a) becomes

$$\text{Convective flux rate (out − in)}_x = \frac{\partial}{\partial x}(u\,C\,A_x)\,dx = \frac{\partial}{\partial x}(u\,C)\,dx\,dy\,dz \qquad (2.10a)$$

Because the normal area, $A_x = dy\,dz$, of our box does not change with x; it can be pulled out of the partial with respect to x. This is done in the second part of equation (2.10a). The same can be done with the y- and z-components of convective flux rate

$$\text{Convective flux rate (out − in)}_y = \frac{\partial}{\partial y}(v\,C\,A_y)\,dy = \frac{\partial}{\partial y}(v\,C)\,dx\,dy\,dz \qquad (2.10b)$$

$$\text{Convective flux rate (out − in)}_z = \frac{\partial}{\partial y}(w\,C\,A_z)\,dz = \frac{\partial}{\partial z}(w\,C)\,dx\,dy\,dz \qquad (2.10c)$$

Finally, adding equations (2.10a) to (2.10c) results in the total net convective flux rate:

$$\text{Net convective flux rate} = \left[\frac{\partial}{\partial x}(u\,C) + \frac{\partial}{\partial y}(v\,C) + \frac{\partial}{\partial z}(w\,C)\right]dx\,dy\,dz \qquad (2.11)$$

Diffusive Flux Rates. For net diffusive flux rate in the x-direction, equation (2.9a) becomes

$$\text{Diffusive flux rate (out − in)}_x = \frac{\partial}{\partial x}\left(-D\frac{\partial C}{\partial x}A_x\right)dx = \frac{\partial}{\partial x}\left(-D\frac{\partial C}{\partial x}\right)dx\,dy\,dz$$
$$(2.12a)$$

A self-test of understanding flux rates would be to look at Figure 2.3 and write out the diffusive flux rates in the y- and z-directions on a separate sheet of paper. The result is similar to equation (2.12a):

$$\text{Diffusive flux rate (out − in)}_y = \frac{\partial}{\partial y}\left(-D\frac{\partial C}{\partial y}A_y\right)dy = \frac{\partial}{\partial y}\left(-D\frac{\partial C}{\partial y}\right)dx\,dy\,dz$$
$$(2.12b)$$

$$\text{Diffusive flux rate (out − in)}_z = \frac{\partial}{\partial z}\left(-D\frac{\partial C}{\partial z}A_z\right)dz = \frac{\partial}{\partial z}\left(-D\frac{\partial C}{\partial z}\right)dx\,dy\,dz$$
$$(2.12c)$$

Finally, we can add equations (2.12a) to (2.12c) to write an equation describing the net diffusive flux rate (out – in) out of the control volume:

$$\text{Net diffusive flux rate} = -\left[\frac{\partial}{\partial x}\left(D\frac{\partial C}{\partial x}\right) + \frac{\partial}{\partial y}\left(D\frac{\partial C}{\partial y}\right) + \frac{\partial}{\partial z}\left(D\frac{\partial C}{\partial z}\right)\right] dx\, dy\, dz$$

$$(2.13)$$

The diffusion coefficient is often not a function of distance, such that equation (2.13) can be further simplified by putting the constant value diffusion coefficient in front of the partial derivative. However, we will also be substituting turbulent diffusion and dispersion coefficients for D when appropriate to certain applications, and they are not always constant in all directions. Therefore, we will leave the diffusion coefficient inside the brackets for now.

Control Volume Mass Balance. We can now combine equations (2.1), (2.5), (2.6), (2.11), and (2.13) into a mass balance on our box for Cartesian coordinates. After dividing by $\overline{V} = dx\, dy\, dz$ and moving the diffusive flux terms to the right-hand side, this mass balance is

$$\frac{\partial C}{\partial t} + \frac{\partial}{\partial x}(u\, C) + \frac{\partial}{\partial y}(v\, C) + \frac{\partial}{\partial z}(w\, C)$$

$$= \left[\frac{\partial}{\partial x}\left(D\frac{\partial C}{\partial x}\right) + \frac{\partial}{\partial y}\left(D\frac{\partial C}{\partial y}\right) + \frac{\partial}{\partial z}\left(D\frac{\partial C}{\partial z}\right)\right] + S \qquad (2.14)$$

When working with a computational transport code, there is little reason to simplify equation (2.14) further. Our primary task, however, is to develop approximate analytical solutions to environmental transport problems, and we will normally be assuming that diffusion coefficient is not a function of position, or x, y, and z. We can also expand the convective transport terms with the chain rule of partial differentiation:

$$\frac{\partial}{\partial x}(u\, C) = u\frac{\partial C}{\partial x} + C\frac{\partial u}{\partial x} \qquad (2.15a)$$

$$\frac{\partial}{\partial y}(v\, C) = v\frac{\partial C}{\partial y} + C\frac{\partial v}{\partial y} \qquad (2.15b)$$

$$\frac{\partial}{\partial z}(w\, C) = w\frac{\partial C}{\partial z} + C\frac{\partial w}{\partial z} \qquad (2.15c)$$

This may not seem like much help, because we have expanded three terms into six. However, if the flow is assumed to be incompressible, a derivation given in fluid mechanics texts (the continuity equation) is

$$\rho\left(\frac{\partial u}{\partial x} + \frac{\partial v}{\partial y} + \frac{\partial w}{\partial z}\right) = 0 \qquad (2.16)$$

where ρ is the density of the fluid. Since equations (2.15a) to (2.15c) are added together in the mass balance equation, the incompressible assumption means that the terms on the far right-hand side of these equations will sum to zero, or

$$\frac{\partial}{\partial x}(u\, C) + \frac{\partial}{\partial y}(v\, C) + \frac{\partial}{\partial z}(w\, C) = u\frac{\partial C}{\partial x} + v\frac{\partial C}{\partial y} + w\frac{\partial C}{\partial z} \qquad (2.17)$$

The incompressible flow assumption is most always accurate for water in environmental applications and is often a good assumption for air. Air flow is close to incompressible as long as the Mach number (flow velocity/speed of sound) is below 0.3. A Mach number of 0.3 corresponds to an air flow velocity of approximately 100 m/s. Equation (2.14) then becomes

$$\frac{\partial C}{\partial t} + u\frac{\partial C}{\partial x} + v\frac{\partial C}{\partial y} + w\frac{\partial C}{\partial z} = D\left(\frac{\partial^2 C}{\partial x^2} + \frac{\partial^2 C}{\partial y^2} + \frac{\partial^2 C}{\partial z^2}\right) + S \tag{2.18}$$

The only assumptions made in developing equation (2.18) are: (1) that diffusion coefficient does not change with spatial coordinate and (2) incompressible flow. We will further simplify equation (2.18) in developing analytical solutions for mass transport problems. In some cases, all we need to do is orient the flow direction so that it corresponds with one of the coordinate axes. We would then have only one convection term.

6. Cylindrical Control Volume

The cylindrical coordinate system and cylindrical control volume are illustrated in Figure 2.6. There are some differences in the development of a mass balance equation on a cylindrical control volume. Primarily, the $rd\theta dx$ side of the control volume increases in area as r increases. For the control volume of Figure 2.6, the area normal to the r-coordinate would be

$$A_r = rd\theta dx \tag{2.19}$$

which is a function of r, one of the independent variables. Then, analogous to equation (2.12), the convective flux in the r-direction would be

$$\text{Convective flux rate (out − in)} = \frac{\partial}{\partial r}\left(v_r\,C\,A_r\right)dr = \frac{\partial}{\partial r}\left(r\,v_r\,C\right)dr\,d\theta\,dz \tag{2.20}$$

The diffusive flux rates would be treated similarly. The area of the control volume changing with radius is the reason the mass transport equation in cylindrical coordinates, given below – with similar assumptions as equation (2.18) – looks somewhat different than in Cartesian coordinates.

$$\frac{\partial C}{\partial t} + \frac{v_r}{r}\frac{\partial}{\partial r}\left(r\,C\right) + \frac{v_\theta}{r}\frac{\partial C}{\partial \theta} + v_z\frac{\partial C}{\partial z} = D\left[\frac{1}{r}\frac{\partial}{\partial r}\left(r\frac{\partial C}{\partial r}\right) + \frac{1}{r^2}\frac{\partial^2 C}{\partial \theta^2} + \frac{\partial^2 C}{\partial z^2}\right] + S \tag{2.21}$$

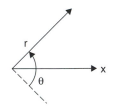

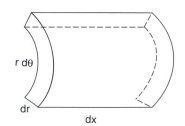

Figure 2.6. Cylindrical control volume.

Derivation of equation (2.21) would help to solidify your understanding of the terms in the diffusion equation.

7. Applications of the Diffusion Equation

We will try our hand at applying the diffusion equation to a couple of mass transport problems. The first is the diffusive transport of oxygen into lake sediments and the use of oxygen by the bacteria to result in a steady-state oxygen concentration profile. The second is an unsteady solution of a spill into the groundwater table.

EXAMPLE 2.1: *Steady oxygen concentration profile in lake sediments (steady-state solution with a first-order sink)*

Given a concentration, C_0, in the overlying water, and a first-order sink of oxygen in the sediments, develop an equation to describe the dissolved oxygen concentration profile in the sediments (see Figure E2.1.1).

Assume:

1. Steady: $\dfrac{\partial}{\partial t} \to 0$
2. No flow: $u, \ v, \ w \to 0$
3. Small horizontal variation: $\dfrac{\partial^2 C}{\partial z^2} >> \dfrac{\partial^2 C}{\partial x^2}, \ \dfrac{\partial^2 C}{\partial y^2}$
4. No sorption: $R = 1$ (accurate for oxygen in sediments)
5. First-order sink: $S = -kc$, where k is a rate constant

Then, the diffusive mass transport equation (2.18) becomes

$$0 = D\frac{\partial^2 C}{\partial z^2} - kC \qquad \text{(E2.1.1)}$$

or, since $C = C(z)$,

$$0 = D\frac{d^2 C}{dz^2} - kC \qquad \text{(E2.1.2)}$$

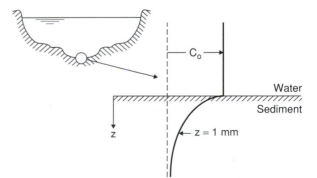

Figure E2.1.1. Illustration of dissolved oxygen profile in lake sediments.

A solution to equation (E2.1.2) requires two boundary conditions because it is a second-order equation. These boundary conditions are:

1. At $z = 0$, $C = C_0$
2. At $z \rightarrow \infty$, $C \Rightarrow 0$

A solution to equation (E2.1.2) may be achieved by (1) separating variables and integrating or (2) solving the equation as a second-order, linear ordinary differential equation. We will use the latter because the solution technique is more general.

1. Assign λ to be the $\dfrac{d}{dz}$ operator. Then, equation (E2.1.2) becomes

$$\left(\lambda^2 - \frac{k}{D} \right) C = 0 \tag{E2.1.3}$$

2. Solve for λ

$$\lambda = \pm \sqrt{k/D}$$

3. The solution, developed in texts on solving ordinary differential equations (Kreysig, 1982), is

$$C = \beta_1 \, e^{\lambda_1 z} + \beta_2 \, e^{\lambda_2 z} \quad \begin{array}{l} \lambda_1 = +\sqrt{k/D} \\ \lambda_2 = -\sqrt{k/D} \end{array} \tag{E2.1.4}$$

4. β_1 and β_2 are determined from boundary conditions:

Apply boundary condition 2 to equation (E2.1.4):

$$C = 0 = \beta_1 \, e^{\sqrt{k/D}\infty} + \beta_2 \, e^{-\sqrt{k/D}\infty}$$

This is only possible if $\beta_1 = 0$. Apply boundary condition 1 to equation (E2.1.4):

$$C_0 = 0 + \beta_2 e^{-0} = \beta_2$$

Thus, the solution is

$$C = C_0 \, e^{\sqrt{k/D}\,z} \tag{E2.1.5}$$

which is plotted in Figure E2.1.2.

At steady state, the oxygen profile is a balance between diffusion from the sediment surface and bacterial use of oxygen in the sediments. If the sediments are mostly sand, the depth of the layer with oxygen can be 10 cm or more. If the sediments have a substantial organic content (like a mud), the aerobic layer (>0.1 g/m^3 oxygen concentration) can be less than 1 mm in depth.

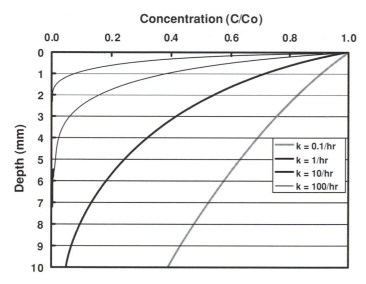

Figure E2.1.2. Solution (equation (E2.1.5)) to Example 2.1. for oxygen concentration in lake sediments with first-order sink.

EXAMPLE 2.2: *Unsteady dissolution of a highly soluble pollutant (herbicides, pesticides, ammonia, alcohols, etc.) into groundwater (unsteady, one-dimensional solution with pulse boundary conditions)*

A tanker truck carrying a highly soluble compound in Mississippi tried to avoid an armadillo at night, ran off the interstate at a high speed, turned over in the drainage

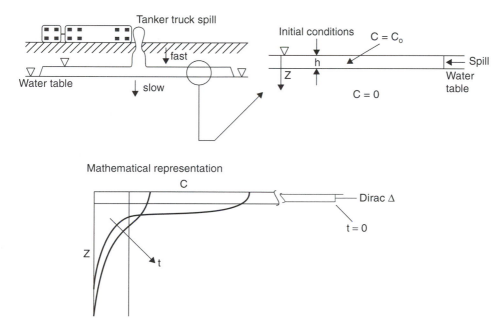

Figure E2.2.1. llustration of the tanker truck spill.

ditch, and spilled a soluble compound. The compound has infiltrated into the ground, and much of it has reached and temporarily spread out over the groundwater table. As part of a spill response team, you need to estimate the groundwater contamination. Predict concentrations over time in the groundwater table. (See Figure E2.2.1.)

The mass transport equation for this example is

$$\frac{\partial C}{\partial t} + u\frac{\partial C}{\partial x} + v\frac{\partial C}{\partial y} + w\frac{\partial C}{\partial z} = D\left(\frac{\partial^2 C}{\partial x^2} + \frac{\partial^2 C}{\partial y^2} + \frac{\partial^2 C}{\partial z^2}\right) + S \qquad (E2.2.1)$$

Assume:

1. Minimal horizontal variations

$$0 \cong \frac{\partial C}{\partial x} = \frac{\partial^2 C}{\partial x^2} \cong \frac{\partial C}{\partial y} \cong \frac{\partial^2 C}{\partial y^2}$$

2. No flow in the vertical direction, $w = 0$
3. No reactions, including adsorption and desorption such that $S = 0$

Then, with these three assumptions, equation (E2.2.1) becomes

$$\frac{\partial C}{\partial t} = D\frac{\partial^2 C}{\partial z^2} \qquad (E2.2.2)$$

We will simulate the initial conditions with these boundary conditions:

1. The mass of chemical is assumed to be spread instantaneously across a very thin layer at $t = 0$ (a Dirac delta in z and t). At $z = 0^+$, $t = 0$, the total mass $= M$; and the total surface area is A.
2. At $z \Rightarrow \infty$, $C \Rightarrow 0$

Equation (E2.2.2), with boundary conditions 1 and 2, has the solution

$$C = \frac{2\,M/A}{\sqrt{4\pi\,Dt}}\, e^{-z^2/4Dt} \qquad (E2.2.3)$$

For the solution of boundary conditions with a Dirac delta, it is easier to provide the solution and then prove that it works. We will do that now.

Equation (E2.2.3) is the correct solution *if* it solves equation (E2.2.2) and meets the boundary conditions. Let's test it:

a) Boundary condition 2:

$$\text{At } z \Rightarrow \infty \qquad e^{-z/4Dt} \Rightarrow 0 \qquad \therefore C \Rightarrow 0$$

Boundary condition 2 is satisfied by equation (E2.2.3).

b) Boundary condition 1:

$$M = A\int_0^\infty C\,dz = A\int_0^\infty \frac{2M/A}{\sqrt{4\pi\,Dt}}\, e^{-z^2/4Dt}\,dz \qquad (E2.2.4)$$

Equation (E2.2.4) looks like the probability integral with the limits given from 0 to ∞. The probability integral is

$$\int_0^\infty e^{-\xi^2}\, d\xi = \frac{\sqrt{\pi}}{2}$$

Now assign $\xi = \dfrac{z}{\sqrt{4Dt}}$; then, $d\xi = \dfrac{dz}{\sqrt{4Dt}}$

or $dz = \sqrt{4Dt}\, d\xi$

and equation (E2.2.4) becomes

$$M = \frac{2M}{\sqrt{\pi}} \int_0^\infty e^{-\xi^2}\, d\xi = \frac{2M}{\sqrt{\pi}} \frac{\sqrt{\pi}}{2} = M \quad \text{!!}$$

Boundary condition 1 is satisfied by the solution:

c) Finally, let us substitute equation (E2.2.3) into (E2.2.2) to see if the solution satisfies our governing equation:

i. $\dfrac{\partial C}{\partial t} = -\dfrac{1}{2}\dfrac{2M/A}{\sqrt{4\pi\, Dt}\,t}\, e^{-z^2/4Dt} + \dfrac{2M/A}{\sqrt{4\pi\, Dt}}\left(\dfrac{z^2}{4Dt^2}\right) e^{-z^2/4Dt}$

or

$$\frac{\partial C}{\partial t} = \frac{-M/A}{\sqrt{4\pi\, Dt^3}}\left(1 - \frac{z^2}{2Dt}\right) e^{-z^2/4Dt} \tag{E2.2.5}$$

ii. $\dfrac{\partial C}{\partial z} = \dfrac{-2z}{4Dt}\dfrac{M/A}{\sqrt{4\pi\, Dt}}\, e^{-z^2/4Dt} = \dfrac{-z\, M/A}{Dt\sqrt{4\pi\, Dt}}\, e^{-z^2/4Dt}$

and

iii. $\dfrac{\partial^2 C}{\partial z^2} = \dfrac{\partial}{\partial z}\left(\dfrac{\partial C}{\partial z}\right) = \dfrac{-M/A}{\sqrt{4\pi\, Dt^3}}\left(\dfrac{1}{D} - \dfrac{z^2}{2D^2 t}\right) e^{-z^2/4Dt}$ $\qquad$ (E2.2.6)

Plug equations (E2.2.5) and (E2.2.6) into equation (E2.2.2)

$$\frac{-M/A}{\sqrt{4\pi\, Dt^3}}\left(1 - \frac{z^2}{2Dt}\right) e^{-z^2/4Dt} = D\left[\frac{-M/A}{\sqrt{4\pi\, Dt^3}}\left(\frac{1}{D} - \frac{z^2}{2D^2 t}\right) e^{-z^2/4Dt}\right]$$

The solution – equation (E2.2.3) – works!

To repeat: equation (E2.2.3) is a solution to the diffusion equation, and we have shown that it meets the boundary conditions. It is therefore a solution to our problem as we have formulated it. This may seem like a fairly extensive example for one solution. However, we can use equation (E2.2.3) as a basis for an entire set of Dirac delta solutions that can model instantaneous spills. Thus, equation (E2.2.3) is a building block for many of the solutions we will model.

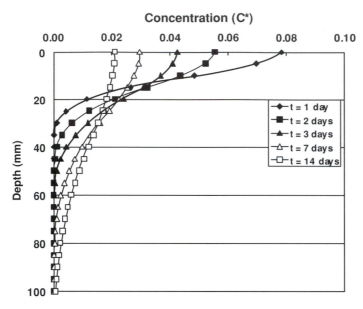

Figure E2.2.2. Solution to the tanker truck spill illustrating groundwater concentration versus distance at various times. $\Delta h = 2\,\mathrm{mm}$, $D = 6 \times 10^{-4}\,\mathrm{mm}^2/\mathrm{s}$.

The solution to equation (E2.2.3) is plotted versus depth at various times in Figure E2.2.2. The solution can also be made dimensionless by assuming that the initial thickness of the spill layer is Δh. Then, a new variable $z = k\,\Delta h$ will be used in assigning

$$\eta = \frac{\Delta h}{\sqrt{4Dt}} \tag{E2.2.7}$$

with

$$C^* = \frac{CA\Delta h}{2M} \tag{E2.2.8}$$

Substituting equation (E2.2.7) and (E2.2.8) into equation (E2.2.3) gives

$$C^* = \frac{\eta}{\sqrt{\pi}}\, e^{-(k\eta)^2} \tag{E2.2.9}$$

The concentration at $z = 0$ decreases as the initial mass is diffused. At low values of time, the concentration at and close to $z = 0$ is strongly dependent on the Δh chosen. At larger times and deeper depths, however, this dependency decreases, and the solution becomes independent of Δh.

It is interesting to note that the solution given as equation is very similar to a Gaussian probability distribution (given in Figure (E2.2.3)), with the following relationship for $P(z)$:

$$P(z) = \frac{1}{\sigma\sqrt{2\pi}} e^{-(z-z_m)^2/2\sigma^2} \tag{E2.2.10}$$

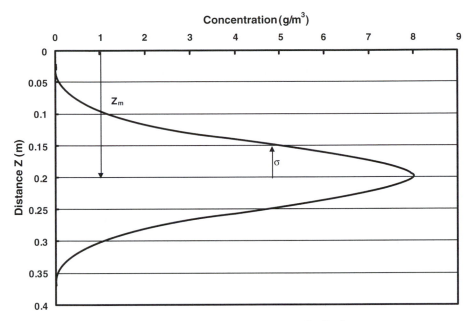

Figure E2.2.3. Concentration profile with depth as a Gaussian distribution.

where z_m is the depth of the maximum concentration (or the center of concentration mass).

Comparing equations (E2.2.3) and (E2.2.7), we can see that

$$2\sigma^2 \Leftrightarrow 4\,Dt$$

or

$$D = \sigma^2 2\,t \qquad (E2.2.11)$$

Note that if we measure σ, we can determine D. The following equations can be used to determine σ (and thus D) from measurements:

Centroid of concentration distribution, z_m (for our problem $\bar{z} = 0$):

$$z_m = \frac{\int\limits_{-\infty}^{\infty} z\,C\,dz}{\int\limits_{-\infty}^{\infty} C\,dz} \cong \frac{\sum (z - z_m)C/\Delta z}{\sum C\,\Delta z} \qquad (E2.2.12)$$

Variance of distribution, σ^2:

$$\sigma^2 = \frac{\int\limits_{-\infty}^{\infty} (z - z_m)^2\,C\,dz}{\int\limits_{-\infty}^{\infty} C\,dz} \cong \frac{\sum (z - z_m)^2 C/\Delta z}{\sum C\,\Delta z} \qquad (E2.2.13)$$

Substituting σ from equation (E2.2.13) into equation (E2.2.11) will give an estimate of D.

B. Adsorption and Desorption in Sediment and Soil

Sorption relates to a compound sticking to the surface of a particle. Adsorption relates to the process of compound attachment to a particle surface, and desorption relates to the process of detachment. Example 2.2 was on a soluble, nonsorptive spill that occurred into the ground and eventually entered the groundwater. We will now review sorption processes because there are many compounds that are sorptive and subject to spills. Then, we can examine the solutions of the diffusion equation as they apply to highly sorptive compounds.

There are many processes involved in determining the chemical thermodynamics (equilibria) of sorption, relating to the polarity and charge of the solute, solvent, and particle. This is not a text on surface chemistry, so we will consider only the properties that get used in the diffusion equation most often. Environmental chemicals are generally classified as hydrophilic (likes water) and hydrophobic (hates water). Water is a polar molecule in that it has two hydrogen atoms on one side and an oxygen atom on the other. Solutes with a polarity or charge, therefore, will have water molecules surrounding them with the tendency to have the proper charge of atom adjacent to the solute. Most amides and alcohols are strongly polar and also soluble in water. These are generally hydrophilic compounds. Other organic compounds with larger molecular weights, especially with aromatic rings, are generally nonpolar and are classified as hydrophobic compounds. It makes sense that these hydrophobic compounds would adsorb to the nonpolar organic material in the sediments or soils. There are handbooks (Lyman et al., 1990) that can be used to estimate the chemical thermodynamics of a water particle system.

How do we handle sorption in our transport equation? For particles that are not transported with the flow field, like sediments and groundwater flow, we are interested in the water concentrations. The sorbed portion of the compound is not in the solute phase and should not be considered in the transport equation, except when transfer of the compound between the water and particles occur. Adsorption would then be a sink of the compound, and desorption would be a source.

Let us assign S_p to be the mass of chemical sorbed to particles per mass of solids contained in our control volume and C to be the concentration of the compound in solution. Then, our source term in the diffusion equation is equal to the rate of change of mass due to adsorption and desorption per unit volume, or

$$S = \frac{\rho_b}{\varepsilon} \frac{\partial S_p}{\partial t} \qquad (2.22)$$

where ρ_b is the bulk density of the solid (mass of solid/volume of fluid and solid), ε is the porosity of the media (volume of fluid/volume of fluid and solid), and $\partial S_p/\partial t$ is the rate of sorption relative to the mass of solid (mass adsorbed/mass of solid/time). If the sorption rate is negative, desorption is occurring. The units of S in equation (2.22) are mass adsorbed/volume of fluid/time. This is similar to the units for the $\partial C/\partial t$ term, which are a change of mass/volume of fluid/time.

The source term in equation (2.22) requires a separate differential equation for S_p, which would incorporate the concentration of the compound in solution. We would thus have two equations that need to be solved simultaneously. However, most sorption rates are high, relative to the transport rates in sediments and soil. Thus, *local equilibrium in adsorption and desorption is often a good assumption.* It also simplifies the solution to a transport problem considerably. If we make that assumption, S_p changes in proportion to C alone, or

$$S_p = S_p(C) \tag{2.23}$$

and

$$\frac{\partial S_p}{\partial t} = \frac{\partial S_p}{\partial C}\frac{\partial C}{\partial t} \tag{2.24}$$

Now, if we substitute equation (2.24) into (2.22), we get

$$S = \frac{\rho_b}{\varepsilon}\frac{\partial Sp}{\partial C}\frac{\partial C}{\partial t} \tag{2.25}$$

The $\partial S_p/\partial C$ term can be found from the equilibrium relationship of Freundlich isotherms, expressed as

$$S_p = K_d C^{\beta} \tag{2.26}$$

where K_d is an equilibrium-partitioning coefficient between the fluid and sorption to the solid, and β is a coefficient fit to measured data. Then,

$$\frac{\partial S_p}{\partial C} = \beta \, K_d \, C^{\beta-1} \tag{2.27}$$

At the lower concentrations normally found in the environment, $\beta = 1$ is a valid assumption. Then, equation (2.27) becomes

$$\frac{\partial S_p}{\partial C} = K_d \, (\beta = 1) \tag{2.28}$$

Substituting equation (2.28) into equation (2.25) now results in a source term that no longer contains the variable S_p and keeps the partial differential equation (PDE) of our mass balance linear:

$$S = \frac{\rho_b}{\varepsilon} K_d \frac{\partial C}{\partial t} \tag{2.29}$$

Now, if we substitute equation (2.29) into our mass transport equation (2.18) for the source term, the result is a PDE where the only dependent variable is C:

$$\frac{\partial C}{\partial t} + u\frac{\partial C}{\partial x} + v\frac{\partial C}{\partial y} + w\frac{\partial C}{\partial z} = D\left(\frac{\partial^2 C}{\partial x^2} + \frac{\partial^2 C}{\partial y^2} + \frac{\partial^2 C}{\partial z^2}\right) - \frac{\rho_b}{\varepsilon} K_d \frac{\partial C}{\partial t} \tag{2.30}$$

or

$$\left(1 + \frac{\rho_b}{\varepsilon} K_d\right)\frac{\partial C}{\partial t} + u\frac{\partial C}{\partial x} + v\frac{\partial C}{\partial y} + w\frac{\partial C}{\partial z} = D\left(\frac{\partial^2 C}{\partial x^2} + \frac{\partial^2 C}{\partial y^2} + \frac{\partial^2 C}{\partial z^2}\right) \tag{2.31}$$

If we divide equation (2.31) by the term $(1 + K_d \rho_b / \varepsilon)$, we can see that all convective and diffusive transport is *retarded* by equilibrium adsorption and desorption. Thus, a *retardation* coefficient is defined:

$$R = \text{retardation coefficient} = 1 + K_d \rho_b / \varepsilon \qquad (2.32)$$

and equation (2.31) becomes

$$\frac{\partial C}{\partial t} + \frac{u}{R}\frac{\partial C}{\partial x} + \frac{v}{R}\frac{\partial C}{\partial y} + \frac{w}{R}\frac{\partial C}{\partial z} = \frac{D}{R}\left(\frac{\partial^2 C}{\partial x^2} + \frac{\partial^2 C}{\partial y^2} + \frac{\partial^2 C}{\partial z^2}\right) \qquad (2.33)$$

Equation (2.33) shows us that we can utilize the retardation coefficient and simply convert all of the transport terms through dividing by R as long as we can assume that the sorption rates are fast compared with our transport rates and the equilibrium partitioning is linearly related to concentration. Thus, if there is a spill into the groundwater table that is highly hydrophobic, it would transport through the soil more slowly than one that is hydrophilic. Both the convective and the diffusive flux would be "retarded" for the hydrophobic compound. If both hydrophylic and hydrophobic compounds are contained in the spill, the hydrophylic compound would show up first at a downstream location. The similarity to the manner in which a chromatographic column separates compounds is not fortuitous, because the column is separating compounds through their sorption to the column's media.

Determination of K_d from Octanol–Water Partitioning Coefficient. There have been a number of empirical equations developed to determine the water–solid partitioning coefficient, K_d (Lyman et al., 1990). These are primarily for the many organic chemicals that exist in the environment, usually due to human impacts. Many of them use the octanol–water partitioning coefficient for the compound as an indicator of hydrophobicity. Octanol is a relatively insoluble organic compound. Because most organic compounds tend to adsorb to the organic portion of the particles, a hydrophobic organic compound placed in an octanol–water solution will tend toward the octanol. The ratio of concentration in the octanol over concentration in the water will indicate the degree of the hydrophobicity. It is a straightforward and relatively easy measurement to make, so most organic compounds of interest in the environment have an octanol–water partitioning coefficient that has been measured.

Karikhoff et al. (1979) developed a simple empirical equation for equilibrium partitioning of organic compounds that will be used in this text (other equations are given in Lyman et al., 1990):

$$K_d = \beta f K_{ow} \qquad (2.34)$$

where K_{ow} is the dimensionless octanol–water partitioning coefficient; f is the fraction of soil that is organic matter (usually from zero in sand to 0.01 in sandy soil to 0.10 in muck); and β is an empirical coefficient, estimated by Karikhoff et al. to be 0.41 cm^3/g. It is generally the organic matter in the medium to which organic compounds adsorb. Hence, the use of organic fraction.

The other parameters required to compute a retardation coefficient are bulk density, ρ_b, and the porosity of the media, ε. The bulk density of the water and soil is typically 1.6 to 2.1 g/cm^3. The porosity of the soil or sediments is typically 0.2 to 0.4. Thus, ρ_b/ε is typically between 4 and 10 g/cm^3.

C. The Product Rule

We will introduce the product rule through demonstrating its use in an example problem. The product rule can be used to expand a solution without source and sink terms to the unsteady, one-dimensional diffusion equation to two and three dimensions. It does not work as well in developing solutions to all problems and therefore is more of a technique rather than a rule. Once again, the final test of any solution is (1) it must solve the governing equation(s) and (2) it must satisfy the boundary conditions.

EXAMPLE 2.3: *Diffusion of a toxic material from a deep burial location under a lake (unsteady, three-dimensional solution to pulse boundary conditions)*

Toxic material was placed many years ago in the sediments of a lake. Sedimentation since that time has continued their burial. Now, a harbor authority would like to put in ship piers with pilings that may disturb the toxic compounds in the sediments. You would like to know if the concentrations of the toxic compound have reached a sufficient level that the pilings will create an environmental threat. Which equation would you use for a quick, first estimate? (See Figure E2.3.1.)

First, write the governing mass transport equation:

$$\frac{\partial C}{\partial t} + \frac{u}{R}\frac{\partial C}{\partial x} + \frac{v}{R}\frac{\partial C}{\partial y} + \frac{w}{R}\frac{\partial C}{\partial z} = \frac{D}{R}\left(\frac{\partial^2 C}{\partial x^2} + \frac{\partial^2 C}{\partial y^2} + \frac{\partial^2 C}{\partial z^2}\right) + S \qquad \text{(E2.3.1)}$$

Assume:

1. No flow: $u = v = w = 0$
2. No reaction except equilibrium sorption: $S \Rightarrow 0$

Then, equation (E2.3.1) becomes

$$\frac{\partial C}{\partial t} = \frac{D}{R}\left(\frac{\partial^2 C}{\partial x^2} + \frac{\partial^2 C}{\partial y^2} + \frac{\partial^2 C}{\partial z^2}\right) \qquad \text{(E2.3.2)}$$

with the following boundary conditions:

1. At $z \to \infty$, $C \to 0$
2. At $y \to \infty$, $C \to 0$
3. At $x \to \infty$, $C \to 0$
4. At $t = 0$, all of the initial mass is located at $(x,y,z) = (0,0,0)$

Therefore, an approximate solution would be to use a Dirac delta with the entire mass of the material, M, located at the origin. (This is a three-dimensional version of

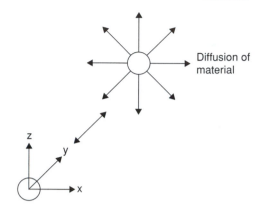

Water

Sediment

Diffusion of
material

Figure E2.3.1. Initial conditions for the diffusion
of toxic material from a barrel.

Example 2.2.) Use the "product rule," which is often applied to expand a PDE solution to three dimensions. Assume that C_1, C_2, and C_3 are one-dimensional solutions to the governing equation with similar boundary conditions. Then, the product rule gives a potential three-dimensional solution of

$$C(x,y,z,t) = C_1(x,t)C_2(y,t)C_3(z,t) \tag{E2.3.3}$$

Substitute equation (E2.3.3) into equation (E2.3.2)

$$C_2C_3\frac{\partial C_1}{\partial t} + C_1C_3\frac{\partial C_2}{\partial t} + C_1C_2\frac{\partial C_3}{\partial t} = \frac{D}{R}\left(C_2C_3\frac{\partial^2 C_1}{\partial x^2} + C_1C_3\frac{\partial^2 C_2}{\partial y^2} + C_1C_2\frac{\partial^2 C_3}{\partial z^2}\right)$$

or

$$C_2C_3\left(\frac{\partial C_1}{\partial t} - \frac{D}{R}\frac{\partial^2 C_1}{\partial x^2}\right) + C_1C_3\left(\frac{\partial C_2}{\partial t} - \frac{D}{R}\frac{\partial^2 C_2}{\partial y^2}\right) + C_1C_2\left(\frac{\partial C_3}{\partial t} - \frac{D}{R}\frac{\partial^2 C_3}{\partial z^2}\right) = 0 \tag{E2.3.4}$$

We will solve equation (E2.3.4) by placing each term in brackets equal to zero. Note that we solved this problem (one bracket solution) in Example 2.2:

$$\left(C = \frac{2M/A}{\sqrt{4\pi Dt}}\,e^{-z^2/4Dt}\right) \tag{E2.3.5}$$

except for in this solution, $D \Rightarrow D/R$ and $2M/A \Rightarrow M$, because diffusion is in both directions, and Example 2.2 was diffusing in one direction only, with zero flux in the upward direction at $z = 0$.

Thus:

$$C_1 = \frac{\beta}{\sqrt{4\pi\,Dt/R}}\,\exp\left(\frac{-Rx^2}{4Dt}\right)$$

$$C_2 = \frac{\beta}{\sqrt{4\pi\,Dt/R}}\,\exp\left(\frac{-Ry^2}{4Dt}\right)$$

$$C_3 = \frac{\beta}{\sqrt{4\pi\,Dt/R}}\,\exp\left(\frac{-Rz^2}{4Dt}\right)$$

β will be determined from boundary conditions. And since $C = C_1\,C_2\,C_3$, equation (E2.2.3) becomes

$$C = \frac{\beta^3}{(4\pi\,Dt/R)^{3/2}}\,\exp\left[\frac{-R(x^2 + y^2 + z^2)}{4Dt}\right] \tag{E2.3.6}$$

The analogy with Example 2.2 indicates that $B = M^{1/3}$ might be a good guess. Let's try this:

$$C = \frac{M}{(4\pi\,Dt/R)^{3/2}}\,\exp\left[\frac{-R(x^2 + y^2 + z^2)}{4Dt}\right] \tag{E2.3.7}$$

Equation (E2.3.7) solves equation (E2.3.2) and satisfies the boundary conditions. Thus, the product rule enabled us to expand our solution to three dimensions.

D. Superposition Principle

The superposition principle can be used to combine solutions for linear partial differential equations, like the diffusion equation. It is stated as follows:

If any two relations solve a given linear partial differential equation, the sum of those two is also a solution to the linear partial differential equation.

To meet a particular application, known solutions of a linear differential equation may be combined to meet the boundary conditions of that application. The superposition principle will be demonstrated through its use in Example 2.4.

EXAMPLE 2.4: *Two leaky barrels of toxic material buried deep under a lake (unsteady, three-dimensional solution with two pulses and superposition)*

Now consider the same situation as in Example 2.3, except that there are two barrels of toxic material buried deep under a lake, separated by a distance Δx, illustrated in Figure E2.4.1. The worst case would be to assume that they began to leak immediately (i.e., that there is no barrier to transport after the barrels are buried).

We will use superposition to adapt our solution of Example 2.3 to this problem. Superposition works for linear PDEs. The governing mass transport equations becomes

$$\frac{\partial C}{\partial t} = \frac{D}{R}\left(\frac{\partial^2 C}{\partial x^2} + \frac{\partial^2 C}{\partial y^2} + \frac{\partial^2 C}{\partial z^2}\right) \tag{E2.4.1}$$

similar to that for Example 2.3. But the boundary conditions are different:

1. At $x = +\infty, -\infty, C = 0$
    $y = +\infty, -\infty, C = 0$
    $z = +\infty, -\infty, C = 0$

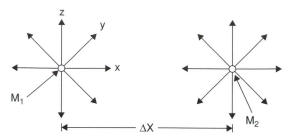

Figure E2.4.1. Boundary conditions for the two leaky barrel problems.

2. At $(x,y,z,t) = (0,0,0,0)$ Dirac delta in C of mass M_1
3. At (x,y,z,t) $(\Delta x,0,0,0)$ Dirac delta C in of mass M_2

Of course, the solution still needs to satisfy the boundary conditions. If $C = C_1$ and $C = C_2$ are solutions to the equation (E2.4.1) for different boundary conditions, then $C = C_1 + C_2$ is also a solution as long as the boundary conditions are satisfied. Thus, Example 2.3 for 1 barrel at $x,y,z = 0,0,0$ may be used to develop a solution for Example 2.4:

For 1 barrel at $x,y,z = 0,0,0$

$$C_1 = \frac{M_1}{\left(4\pi \frac{D}{R}t\right)^{3/2}} \exp\left[-\frac{R\left(x^2 + y^2 + z^2\right)}{4Dt}\right] \tag{E2.4.2}$$

For 1 barrel at $x,y,z = \Delta x,0,0$

$$C_2 = \frac{M_2}{\left(4\pi \frac{D}{R}t\right)^{3/2}} \exp\left\{-\frac{R\left[(x - \Delta x)^2 + y^2 + z^2\right]}{4Dt}\right\} \tag{E2.4.3}$$

Thus, applying the superposition principle

$$C = C_1 + C_2 \tag{E2.4.4}$$

or

$$C = \frac{M_1}{(4\pi\ Dt/R)^{3/2}} \exp\left[\frac{-R(x^2 + y^2 + z^2)}{4Dt}\right]$$
$$+ \frac{M_2}{(4\pi\ Dt/R)^{3/2}} \exp\left\{\frac{-R[(x - \Delta x)^2 + y^2 + z]^2}{4Dt}\right\} \tag{E2.4.5}$$

Equation (E2.4.5) solves the PDE of equation (E2.4.1) and meets the boundary conditions. Here's how:

Substitute equation (E2.4.4) into equation (E2.4.1):

$$\frac{\partial(C_1 + C_2)}{\partial t} = \frac{D}{R}\left[\frac{\partial^2(C_1 + C_2)}{\partial x^2} + \frac{\partial^2(C_1 + C_2)}{\partial y^2} + \frac{\partial^2(C_1 + C_2)}{\partial z^2}\right] \tag{E2.4.6}$$

The individual terms of equation (E2.4.6) become

$$\frac{\partial(C_1 + C_2)}{\partial t} = \frac{\partial C_1}{\partial t} + \frac{\partial C_2}{\partial t}$$

$$\frac{\partial^2(C_1 + C_2)}{\partial x^2} = \frac{\partial^2 C_1}{\partial x^2} + \frac{\partial^2 C_2}{\partial x^2}$$

and so forth for the y- and z-directions.

Then equation (E2.4.6) becomes

$$\frac{\partial C_1}{\partial t} \qquad \frac{D}{R}\left(\frac{\partial^2 C_1}{\partial x^2} + \frac{\partial^2 C_1}{\partial y^2} + \frac{\partial^2 C_1}{\partial z^2}\right)$$

$$+ \qquad = \qquad +$$

$$\frac{\partial C_2}{\partial t} \qquad \frac{D}{R}\left(\frac{\partial^2 C_2}{\partial x^2} + \frac{\partial^2 C_2}{\partial y^2} + \frac{\partial^2 C_2}{\partial z^2}\right)$$

which can be seen as two equations similar to equation (E2.3.2) Thus, each solution for C_1 and C_2 solves a partial differential equation similar to Example 2.3.

1. Image Solutions to Simulate Boundary Conditions

A superposition "image" can also be used at no-flux and high-flux boundaries (as shown in Figures 2.7 and 2.8). A no-flux boundary can be modeled by a similar image on the other side of the boundary. A high-flux boundary can be modeled with an image of equal but opposite strength (a negative image) on the other side of the boundary.

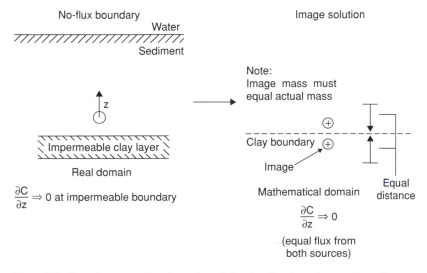

Figure 2.7. Use of superposition image to satisfy a low-flux boundary, such as a liner.

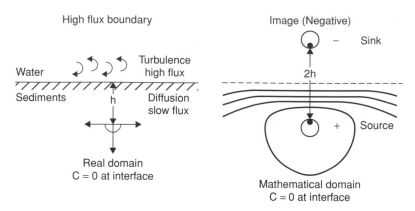

Figure 2.8. Use of a negative image to satisfy a high-flux boundary, such as might occur at an interface.

2. Superposition in Time

A superposition of time is also possible, because, mathematically, time is simply the fourth dimension. Thus, if two solutions at different times are each solutions to the governing linear differential equation, they can be added together to get a superimposed solution that will meet the boundary conditions of the application. This is demonstrated in Example 2.5.

EXAMPLE 2.5: *Two barrels of toxic material buried deep under a lake that begin to leak at different times (unsteady, three-dimensional solution with two pulses and superposition of time and space)*

Now consider the same situation as in Example 2.4, except that there the two barrels of toxic material buried deep under a lake that begin to leak at different times. We will *superimpose time* as well as *distance* to solve this example.

The governing mass transport equation is the same as Example 2.4:

$$\frac{\partial C}{\partial t} = \frac{D}{R}\left(\frac{\partial^2 C}{\partial x^2} + \frac{\partial^2 C}{\partial y^2} + \frac{\partial^2 C}{\partial z^2}\right) \tag{E2.5.1}$$

But the boundary conditions are different:

1. At $x = +\infty, -\infty, C = 0$
    $y = +\infty, -\infty, C = 0$
    $z = +\infty, -\infty, C = 0$
2. At $(x,y,z,t) = (0,0,0,0)$ Dirac delta in C of mass M_1
3. At $(x,y,z,t) = (\Delta x, 0, 0, t_1)$ Dirac delta C in of mass M_2

We will still solve equation (E2.5.1) as the summation of two separate solutions, but with one of them shifted in time (i.e., $t^* = t - t_1$):

For one barrel at $x, y, z = 0, 0, 0$

$$C_1 = \frac{M_1}{\left(4\pi \frac{D}{R} t\right)^{3/2}} \exp\left[-\frac{R(x^2 + y^2 + z^2)}{4Dt}\right] \tag{E2.5.2}$$

For one barrel at $x, y, z = \Delta x, 0, 0$

$$C_2 = \frac{M_2}{\left[4\pi \frac{D}{R}(t - t_1)\right]^{3/2}} \exp\left\{-\frac{R[(x - \Delta x)^2 + y^2 + z^2]}{4Dt}\right\} \tag{E2.5.3}$$

Again, applying the superposition principle

$$C = C_1 + C_2 \tag{E2.5.4}$$

or

$$C = \frac{M_1}{(4\pi\, Dt/R)^{3/2}} \exp\left[-\frac{R(x^2 + y^2 + z^2)}{4Dt}\right]$$

$$+ \frac{M_2}{[4\pi\, D(t - t_1)/R]^{3/2}} \exp\left\{-\frac{R[(x - \Delta x)^2 + y^2 + z]^2}{4D(t - t_1)}\right\} \tag{E2.5.5}$$

Of course, the solution still needs to satisfy the boundary conditions. As Δx becomes small, equation (E2.5.5) becomes a solution for two pulses at different times from one location.

3. Solution to the Diffusion Equation with a Slow Release of Chemical

Equation (E2.5.5) will be used to demonstrate the solution to a continuous source of mass, which can be visualized by considering a large number of small pulses, which can be integrated to result in the concentration profile over time. This is done in Example 2.6.

EXAMPLE 2.6: *One barrel of toxic material buried deep under a lake that begins to leak continuously over time (unsteady, three-dimensional solution with a continuous source of mass)*

The governing mass transport equation is again the same as Example 2.4:

$$\frac{\partial C}{\partial t} = \frac{D}{R}\left(\frac{\partial^2 C}{\partial x^2} + \frac{\partial^2 C}{\partial y^2} + \frac{\partial^2 C}{\partial z^2}\right) \tag{E2.6.1}$$

But the boundary conditions are different:

1. At $x = +\infty, -\infty, C = 0$
   $y = +\infty, -\infty, C = 0$
   $z = +\infty, -\infty, C = 0$
2. At $(x, y, z, t) = (0, 0, 0, t)$ source of mass at rate $\dot{M}$(g/s)

The solution to this problem will be found by adding a large number of individual pulses that occur at $(x,y,z) = (0,0,0)$ and are separated by a short time difference, $d\tau$. Looking at equation (E2.5.5), we can see that this would be accomplished by the equation

$$C = \int_0^t \frac{\dot{M}}{(4\pi D(t-\tau)/R)^{3/2}} \exp\left[-\frac{R(x^2 + y^2 + z^2)}{4D(t-\tau)} \right] d\tau \qquad (E2.6.2)$$

The source of mass, $\dot{M}$, could be a function of τ, or could be a steady value. The solution to equation (E2.6.2) is normally found with a numerical integration routine using integration techniques such as Simpson's rule.

E. Solution to the Diffusion Equation with a Step in Concentration

Sometimes the boundary conditions can be approximated as a step in concentration instead of a step in mass released. This subtle difference in boundary conditions changes the solution from one that is related to pulse boundaries (known mass release) to one resulting from a concentration front with a known concentration at one boundary.

EXAMPLE 2.7: *Dichlorobenzene concentration in lake sediments due to a plating facility discharge (solution to a concentration front)*

For many years, a plating facility for a telecommunications company let their rinse waters flow into an adjacent lake. The compounds used in their rinse included dichlorobenzene, which is a semivolatile compound that also has a fairly high tendency to adsorb to organic compounds in the sediments. Within a few years of the plating facility opening, the dichlorobenzene concentration reached a steady-state value in the lake waters as illustrated in Figure E2.7.1. Estimate the buildup of dichlorobenzene in the sediments during the 50 years since the facility opened until it stopped discharging its untreated waste water.

Assume:

1. Biodegradation is small. $S \to 0$, except for sorption.
2. Variation in x and y are small

$$\frac{\partial^2 C}{\partial x^2}, \quad \frac{\partial^2 C}{\partial y^2} \ll \frac{\partial^2 C}{\partial z^2}$$

3. No flow in sediments under the lake: $u = v = w = 0$
4. $D \cong 6 \times 10^{-10} \text{m}^2/\text{s}$
5. $\dfrac{\rho_b}{\varepsilon} = 6.3, \quad K_d = 15 \text{ cm}^3/\text{g} \quad \Rightarrow \quad R = 1 + \dfrac{\rho_b}{\varepsilon} K_d = 96$

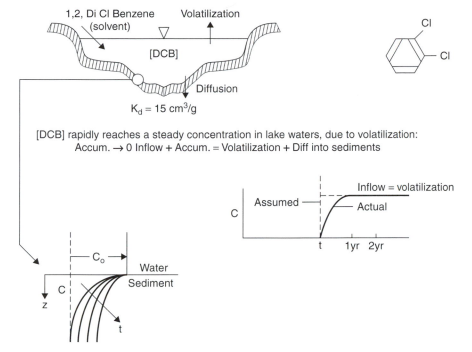

Figure E2.7.1. Illustration of a concentration front moving down into the sediments of a lake.

Then, the diffusion equation for the sediments becomes

$$\frac{\partial C}{\partial t} = \frac{D}{R} \frac{\partial^2 C}{\partial z^2} \qquad (E2.7.1)$$

with boundary conditions:

1. $t > 0, z = 0; C = C_0$
2. $t = 0, z \neq 0; C = 0$

There are three known techniques to solve equation (E2.7.1): Laplace transforms, Fourier transforms, and change of variables, which incorporates both luck and skill. We will use change of variables:

$$\text{Assign:} \quad \eta = \frac{z}{\sqrt{4Dt/R}}$$

$$\frac{\partial C}{\partial t} = \frac{\partial C}{\partial \eta} \frac{\partial \eta}{\partial t} = \frac{-1}{4} \frac{z}{\sqrt{Dt/Rt}} \frac{\partial C}{\partial \eta} = \frac{-\eta}{2t} \frac{\partial C}{\partial \eta}$$

$$\frac{\partial C}{\partial z} = \frac{\partial C}{\partial \eta} \frac{\partial \eta}{\partial z} = \frac{1}{2\sqrt{Dt/R}} \frac{\partial C}{\partial \eta}$$

$$\frac{\partial^2 C}{\partial z^2} = \frac{\partial}{\partial \eta} \left(\frac{\partial C}{\partial z} \right) \frac{\partial \eta}{\partial z} = \frac{R}{4Dt} \frac{\partial^2 C}{\partial \eta^2}$$

Then, equation (E2.7.1) becomes

$$\frac{-\eta}{2t} \frac{\partial C}{\partial \eta} + \frac{D}{R} \left(\frac{R}{4Dt} \frac{\partial^2 C}{\partial \eta^2} \right) = 0$$

or

$$\frac{d^2C}{d\eta^2} + 2\eta\frac{dC}{d\eta} = 0 \tag{E2.7.2}$$

Equation (E2.7.2) may be written

$$\frac{dC'}{d\eta} + 2\eta C' = 0 \quad \text{where} \quad C' = \frac{dC}{d\eta}$$

or

$$\frac{1}{C'}dC' = -2\eta\,d\eta \tag{E2.7.3}$$

We can integrate equation (E2.7.3)

$$\ln C' = -\eta^2 + \beta_0$$

or

$$C' = e^{\beta_0}\,e^{-\eta^2} = \beta_1\,e^{-\eta^2}$$

Now integrate again

$$C = \beta_1 \int_0^\eta e^{-\eta^2}\,d\eta + \beta_2 \Rightarrow \beta_1 \int_0^\eta e^{-\phi^2}\,d\phi + \beta_2 \tag{E2.7.4}$$

Now, note that the error function is given as

$$\text{erf}(\eta) = \frac{2}{\sqrt{\pi}} \int_0^\eta e^{-\phi^2}\,d\phi$$

and the complementary error function is $\text{erfc}(\eta) = 1 - \text{erf}(\eta)$. Values of the error function and complimentary error function for various values of η are given in Appendix A–5. The error function is designed such that $\text{erf}(\infty) = 1$, $\text{erfc}(\infty) = 0$, $\text{erf}(0) = 0$, and $\text{erfc}(0) = 1$. Equation (E2.7.4) may therefore be written as

$$C = \beta_1\text{erf}(\eta) + \beta_2 \tag{E2.7.5}$$

Now we need to determine our boundary conditions in terms of η:

1. $t > 0, z = 0$ $\eta = 0$ $C = C_0$
2. $t = 0, z = 0$ $\eta = \infty$ $C = 0$

Checking other boundary conditions:

$$t \rightarrow \infty \quad \eta \Rightarrow 0 \quad C = C_0$$
$$z \rightarrow \infty \quad \eta \Rightarrow \infty \quad C = 0$$

Now, at $\eta = 0$, $C = C_0$, thus

$$C_0 = \beta_1 0 + \beta_2$$

or

$$\beta_2 = C_0$$

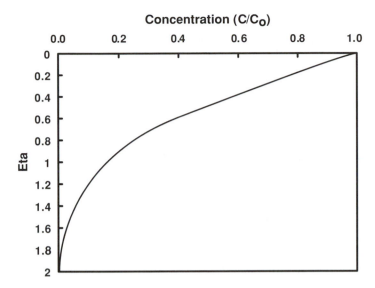

Figure E2.7.2. Illustration of the effect of the change of variables used in Example 2.7.

At $\eta = \infty$, $C = 0$

$$0 = \beta_1 1 + C_0$$

or

$$\beta_1 = -C_0$$

Then, our solution to equation (E2.7.2) is

$$C = C_0 \left[1 - \text{erf} \left(\frac{z}{\sqrt{4Dt/R}} \right) \right] = C_0 \, \text{erfc} \, (\eta) \tag{E2.7.6}$$

which is illustrated in Figure E2.7.2.

We will now apply equation (E2.7.6) to estimate the dichlorobenzene penetration versus time from spillage. The results are given in Table E2.7.1, which gives the interstitial dichlorobenzene concentrations.

Total concentrations (TDCB(z)) include compound adsorbed to the sediments.

$$\text{TDCB}(z) = C(z) + (1 - \varepsilon)\rho_s S \tag{E2.7.7}$$

where

$\varepsilon = $ porosity

$1 - \varepsilon = $ % by volume sediment $\cong 0.6$

$\rho_s = $ density of sediment $\cong 2.5 \text{ g/cm}^3$

$S = $ concentration of sorbed compound (grams of dichlorobenzene/grams of sediment)

Since

$$S = K_d C$$

Table E2.7.1: *Penetration of dichlorobenzene into the sediment over time*

Time, z	1 yrs, C/C_0	4 yrs, C/C_0	10 yrs, C/C_0	50 yrs, C/C_0
1 mm	0.96	0.98	0.988	0.994
1 cm	0.62	0.803	0.87	0.94
10 cm	0	0.015	0.11	0.48
20 cm	0	0	0.01	0.16
30 cm	0	0	0	0.03
100 cm	0	0	0	0

Equation (E2.7.7) becomes

$$\text{TDCB}(z) = C(z)[1 + (1-\varepsilon)\rho_s K_d] = C(z)[1 + 0.6(2.5 \text{ g/cm}^3)(15 \text{ cm}^3/\text{g})]$$

or

$$\text{TDCB} = 23.5 \, C(z)$$

Thus, the total dichlorobenzene per volume of sediment and water would be 23.5 times the concentrations given in Table E2.7.1.

F. Solutions with Reactions

EXAMPLE 2.8: *Remediation of sediments contaminated by radioactivity (unsteady, one-dimensional solution to pulse boundary conditions with reaction)*

Sediments in the Mississippi River were accidentally contaminated with a low-level radioactive waste material that leaked from a nuclear power plant on the river. Pore water concentrations of radioactive compounds were measured following the spill and found to be 10^{-7}g/m^3 over a 2-mm depth. The water contamination was 30% radioactive cesium (^{137}Cs), with a half-life of 30 years, and 70% radioactive cobalt (^{60}Co), with a half-life of 6 years. Objections by the local residents are prevent-ing clean-up efforts because some professor at the local state university convinced them that dredging the sediments and placing them in a disposal facility downstream would expose the residents to still more radioactivity. The state has decided that the sediments should be capped with 10 cm of clay and needs a quick estimate of the diffusion of radioactive material through the clay cap (Figure E2.8.1). If the drinking water limit (10^{-9} g/m^3) is reached at mid-depth in the cap, the state will increase its thickness. Will this occur?

Relevant Data. For radioactive cesium and cobalt, $K_d = 7 \times 10^7$ and 10^7, respec-tively. Assume $\rho_b/\varepsilon = 6$ g/cm^3 for clay under low compressive force. The colloidal concentration in the interstitial water is 10 mg/m^3 (μg/L). The following assumptions will help to achieve our estimate:

1. The contaminant layer is very thin, such that a Dirac delta pulse is applicable.

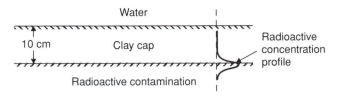

Figure E2.8.1. Illustration of clay cap over sediments contaminated with radioactive material.

2. The colloidal material (very small particles) in the water has a similar K_d to the clay.
3. The effective diffusivity of colloids carrying a radioactive element through clay is $10^{-10} \, \text{m}^2/\text{s}$.
4. The effective diffusivity of the radioactive elements is $6 \times 10^{-10} \, \text{m}^2/\text{s}$.

Then, the mass of each radioactive element is found from the pore-water measurements (primes signify radioactive elements adsorbed to colloids in the pore water):

$$^{137}\text{Cesium} = M_1/A = (C_1 + C_1')\Delta z = 0.3(10^{-7} \, \text{g/m}^3)(0.005 \, \text{m}) = 15 \times 10^{-10} \, \text{g/m}^2$$

$$^{60}\text{Cobalt} = M_2/A = (C_2 + C_2')\Delta z = 0.7(10^{-7} \, \text{g/m}^3)(0.005 \, \text{m}) = 3.5 \times 10^{-10} \, \text{g/m}^2$$

where A is the surface area of the spill.

Coefficients. The retardation coefficients are

$$R_1 = 1 + 6 \, \text{g/cm}^3 (7 \times 10^7 \text{cm}^3/\text{g}) = 4.2 \times 10^7$$
$$R_2 = 1 + 6 \text{g/cm}^3 (10^7 \text{cm}^3) = 6 \times 10^6$$

In addition, the first-order reaction rate coefficients for each radioactive element are found the half-lives, assuming a first-order reaction. For a first-order reaction occurring in a well-mixed container,

$$\frac{C}{C_i} = e^{-kt} \qquad (E2.8.1)$$

where C_i is the concentration at $t = 0$ and k is the reaction rate coefficient. The concentration is equal to half of C_i at the reaction half-life, or from equation (E2.8.1):

$$0.5 = e^{-kt_{1/2}} \qquad (E2.8.2)$$

Equation (E2.8.2) gives

$$k = -\frac{1}{t_{1/2}}\ln(0.5) = \frac{0.69}{t_{1/2}}$$

Then, for our two radioactive elements:

$$k_1 = \frac{0.69}{30 \, \text{yrs}(365 \, \text{days/yr})(24 \, \text{hrs/day})(3600 \, \text{s/hr})} = 7.3 \times 10^{-10} \text{s}^{-1}$$

and

$$k_2 = \frac{0.69}{6\,\text{yrs}(365\,\text{days/yr})(24\,\text{hrs/day})(3600\,\text{s/hr})} = 3.6 \times 10^{-9}\text{s}^{-1}$$

Colloids vs. Free Element. The concentration ratio of colloids with an adsorbed radioactive element to free element is

$$\frac{C_1'}{C_1} = \frac{C_c\,K_d C_1}{C_1} = 10^{-2}\text{g/cm}^3(7 \times 10^6\,\text{cm}^3/\text{g}) = 7 \times 10^4$$

Thus, the radioactive element in the pore water is essentially all associated with the colloids (i.e., $C_1 + C_1' \cong C_1'$ and $C_2 + C_2' \cong C_2'$).

Now, because the water-borne radioactive element is predominantly associated with the colloids, we no longer have a need for the distribution coefficient. There will still be a partitioning because the major portion of the radioactive elements will still be adsorbed to the sediment. This is a separate equilibrium partitioning coefficient, requiring a new experiment on the clay sediments and the colloids present. The partitioning colloid-clay ratio would most likely be dependent on the surface areas of each present in the sediments. A separate size distribution analysis has resulted in a sediment-colloid surface area ratio of 99:1 for the sediment. This results in a "colloid" retardation coefficient of $R_c = 100$ rather than $R_1 = 4.2 \times 10^7$ or $R_2 = 6 \times 10^6$.

Mass Transport Equation. Because the radioactivity will decay regardless of whether the radioactive element is in the free state or adsorbed to clay or colloid, the diffusion equations may be written as

$$\frac{\partial C_1'}{\partial t} = \frac{D_1}{R_c}\frac{\partial^2 C_1'}{\partial z^2} - k_1\,C_1' \tag{E2.8.3}$$

and

$$\frac{\partial C_2'}{\partial t} = \frac{D_2}{R_c}\frac{\partial^2 C_2'}{\partial z^2} - k_2\,C_2' \tag{E2.8.4}$$

Note that the reaction rate coefficients are not divided by R, because radioactive decay will occur whether or not the elements are adsorbed to the sediments. With the exception of the source/sink terms, these equations look like that provided in Example 2.2. The solution will be simply given here with the reader required to apply a technique similar to Example 2.2 that shows that the solution is correct.

$$C_1' = \frac{C_{1i}\Delta z}{\sqrt{4\pi\,Dt/R_c}}\,\exp\left(\frac{-z^2 R_c}{4Dt} - k_1 t\right) \tag{E2.8.5}$$

and

$$C_2' = \frac{C_{2i}\Delta z}{\sqrt{4\pi\,Dt/R_c}}\,\exp\left(\frac{-z^2 R_c}{4Dt} - k_2 t\right) \tag{E2.8.6}$$

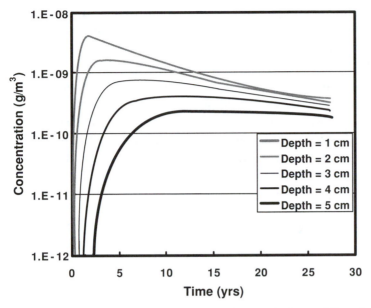

Figure E2.8.2. Radioactive contamination at various depths from spill into clay cap.

where $C_{1i} = 0.3 \times 10^{-7} \text{g/m}^3$, $C_{2i} = 0.7 \times 10^{-7} \text{g/m}^3$, $D = 10^{-10} \text{ m}^2/\text{s}$, $\Delta z = 0.002$ m, $R_C = 100$, $k_1 = 7.3 \times 10^{-10} \text{s}^{-1}$, and $k_2 = 3.6 \times 10^{-9} \text{s}^{-1}$.

The results for total radioactive concentration $C = C_1' + C_2'$ at 5 cm depth are given in Figure E2.8.2. The estimates indicate that at $z = 5$ cm, the maximum concentration will be 2.35×10^{-10} g/m^3 at 15 years after the spill. For these conditions, the state does not need to increase the thickness of the cap.

Linearized Reaction Terms

A linear PDE, such as the diffusion equation, can only have first-order and zero-order source or sink rates. But what if the source or sink term is of higher order? An example would be the generalized reaction

$$C + B \overset{k_1}{\Leftrightarrow} E \tag{2.35}$$

with a forward reaction rate constant of k_1 and an equilibrium constant of

$$K_e = [C][B]/[E] \quad \text{at equilibrium} \tag{2.36}$$

Then,

$$S = k_1(-[C][B] + K_e[E]) \tag{2.37}$$

with two independent concentrations, [B] and [E]. To solve this problem precisely, we need to first solve similar PDEs for E and B. The alternative is that we could use a computational routine and solve for all three concentrations simultaneously.

There is another option, however, that we should consider, even if we are going to use a computational routine. The compound of interest in environmental applications is often a pollutant with low concentrations relative to the other reactants and products. Consider the situation where [B] = 1 ppm, [E] = 1 ppm, [C] = 1 ppb, and $K_e = 10^{-4}$. Then, if [C] is reduced by 0.1 ppb or 10% due to the reaction given in Equation (2.35), [B] would be decreased by 0.1 ppb as well, or 0.01%, and [E] would increase by 0.01%. In this case, a 10% change in the concentration of compound C would only affect the concentration of compounds B and E by 0.01%. The concentration of compounds B and E would be virtually unaffected by the reaction. Since compounds B and E would be very close to constant, why not consider them as a constant? Then, equation (2.37) becomes

$$S = k_2(C - C_e) \tag{2.38}$$

with

$$k_2 = k_1[B] \cong \text{constant} \tag{2.39}$$

and

$$C_e = K_e[E]/[B] \cong \text{constant} \tag{2.40}$$

Equation (2.38) has a first-order sink and a zero-order source, which meets our criteria for an analytical solution to the diffusion equation. C_e is the concentration of C at equilibrium for the reaction. This technique of assuming that multiple reactions are zero-order and first-order reactions will be utilized in Example 2.9.

EXAMPLE 2.9: *Degradation of 1,3-butadiene resulting from a spill (linearized source-sink terms)*

On June 30, 1992, a train derailed over the Nemadji River near Superior, Wisconsin, at 3:00 A.M. The train car that fell into the river was carrying a hydrocarbon-based solvent that included 2% by weight 1,3-butadiene. The butadiene evaporated quickly, and created a "blue cloud" that hung over the river until the wind picked it up at about 7:00 A.M. and created something of a panic in Superior and Duluth, Minnesota. You are part of a forensic engineering investigation to determine what the cloud was and assign the source/sink terms to be used in the transport equations.

$$C_4H_6 \qquad + ROH \xrightarrow{k_1} \quad C_3H_4O + RCH_3$$

1,3-Butadiene   **Hydroxyl**   **Acrolein**
                       Radical

$$C_3H_4O \qquad + ROH \xrightarrow{k_2} \quad 2CH_2O + RCH$$

Acrolein   **Hydroxyl**   **Formaldehyde**
                       Radical

Then,

$$S(C_4H_6) = - k_1[ROH][C_4H_6]$$
$$S(C_3H_6O) = - k_1[ROH][C_4H_6] - k_2[ROH][C_3H_4O]$$
$$S(CH_2O) = k_2[C_3H_4O][ROH]$$

Let's consider the concentrations of ROH, C_4H_6, and so on:

$$[ROH] \sim 1 \text{ ppm(v)}$$
$$[C_4H_6]_{max} \sim 10 \text{ ppb(v)}$$

If all of the C_4H_6 reacts, [ROH] is reduced by only 1%. Therefore, since

$$[C_3H_4] < 10 \text{ ppb(v)}$$

and

$$[CH_2O] < 10 \text{ ppb(v)}$$

assume [ROH] is constant

$$S_{C_4H_6} = - k_1[ROH][C_4H_6] = - k_1'[C_4H_6] \qquad (E2.9.1)$$

$$S_{C_3H_6O} = - k_1'[C_4H_6] - k_2'[C_3H_4O] \qquad (E2.9.2)$$

$$S_{CH_2O} = k_2' \, [C_3H_4O] \qquad (E2.9.3)$$

where $k_2' = k_2[ROH]$.

Then, this equation ($C = [C_4H_6]$) may be used to estimate 1,3-butadiene (C_b) concentrations:

$$\frac{\partial C_b}{\partial t} + U\frac{\partial C_b}{\partial x} = D_x\frac{\partial^2 C_b}{\partial x^2} + Dy\frac{\partial^2 C_b}{\partial x^2} + D_z\frac{\partial^2 C_b}{\partial z^2} - k_1' C_b \qquad (E2.9.4)$$

to estimate acrolein concentrations (C_a):

$$\frac{\partial C_a}{\partial t} + U\frac{\partial C_a}{\partial x} = D_x\frac{\partial^2 C_a}{\partial x^2} + D_y\frac{\partial^2 C_a}{\partial y^2} + D_z\frac{\partial^2 C_a}{\partial z^2} + k_1' C_b - k_2' C_a \qquad (E2.9.5)$$

and to estimate formaldehyde concentrations (C_f):

$$\frac{\partial C_f}{\partial t} + U\frac{\partial C_f}{\partial x} = D_x\frac{\partial^2 C_f}{\partial x^2} + D_y\frac{\partial^2 C_f}{\partial y^2} + D_z\frac{\partial^2 C_f}{\partial z^2} + k_2' C_a \qquad (E2.9.6)$$

The boundary conditions and rate constants can then be used to solve equations (E2.9.4) through (E2.9.6) for the three concentrations. The solution to these three equations would require estimates of wind conditions and release rates from the slick. All that we know is that conditions were "calm" but not so calm that diffusive transport was dominant.

Discussions with eyewitnesses, however, have been compiled and used to estimate that the cloud was 100 m high, 100 m wide, and 15 km long at 7:00 A.M., when the wind began moving the cloud. The train car lost 1,000 kg of solvent before 7:00 A.M. If we assume that the 1,3-butadiene evaporated into a chamber that is 100 m × 100 m × 15,000 m, and relatively well-mixed, we have an initial concentration of

$$C_{b0} = \frac{1,000,000\,\text{g}\,(0.02)}{(100)(100)(15,000)\,\text{m}^3} = 0.00014\,\text{g/m}^3$$

$$= \frac{29\,\text{g/mole}}{54\,\text{g/mole}} \left(\frac{0.00014\,\text{g/m}^3}{1.2\,\text{g/m}^3} \right) = 76\,\text{ppm}$$

where the density of air is approximately 1.2 g/m^3. In addition, if we assume that the "chamber" was well mixed, the transport terms drop out of equation (E2.9.4) through (E2.9.6). Then, equation (E2.9.4) becomes

$$\frac{dC_b}{dt} = -k_1' C_b \tag{E2.9.7}$$

and

$$\frac{dC_a}{dt} = -k_1' C_b - k_2' C_a \tag{E2.9.8}$$

$$\frac{dC_f}{dt} = -k_a' C_a \tag{E2.9.9}$$

In that region, a typical hydroxyl radical concentration would be 1×10^4 ppm. In addition, the rate of reaction coefficients are $k_1 = 3.83 \times 10^{-8}$(ppm s)$^{-1}$ and $k_2 = 0.96 \times 10^{-8}$(ppm s)$^{-1}$.

Then, equation (E2.9.7) gives

$$C_b = C_{b0}\, e^{-k_1' t} = 7,160\,\text{ppm}\, e^{-3.8 \times 10^{-4} t} \tag{E2.9.10}$$

Equation (E2.9.8) gives

$$C_a = C_{b0} \left(e^{-2' t} - e^{-k_1' t} \right) \tag{E2.9.11}$$

and finally equation (E2.9.9) gives

$$C_f = C_{b0} \left[1 - e^{-k_2' t} - \frac{k_2'}{k_1'} \left(1 - e^{-k_1' t} \right) \right] \tag{E2.9.12}$$

The solution to these three equations, plotted in Figure E2.9.1, indicate that formaldehyde was the principal contaminant in the cloud by 7:00 A.M. and is probably what gave the cloud its blue color.

G. Problems

1. Derive the diffusion equation for cylindrical coordinates.

2. Typical values of sediment oxygen demand (SOD) in lakes vary from 0.5 to 5 g/m^2-day. Assuming a diffusivity of 10^{-9} m^2/s for oxygen in the lake sediments,

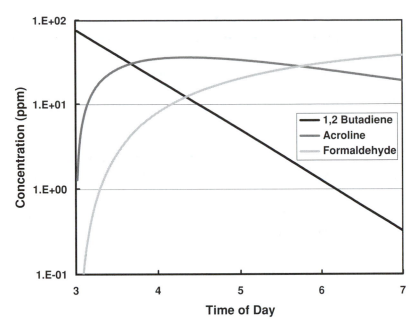

Figure E2.9.1. Concentration in the contaminant cloud.

and an oxygen concentration in the water above the sediments of 10 ppm, determine the range of bacterial oxygen sink coefficients, k, implied by this range of SOD. Determine and plot the steady-state oxygen concentration in the sediments of the lake for the high and low k-values.

3. Monod kinetics are often a better description of the reactions occurring in sediment. Assume that the flux conditions are similar to problem 2 for SOD. The maximum bacterial growth rate is $0.5 \ \mathrm{hr}^{-1}$ in these sediments, and the half-saturation coefficient is $1 \ \mathrm{g/m^3}$ of oxygen. What is the microbacterial population for the 0.5- and 5-g/m²-day cases? (*Hint*: use a change of variables to solve for $f = dC/dz$.) Plot oxygen concentration for each case on the same scale as for problem 2 and compare the two plots.

4. Hexachlorobenzene (C_6Cl_6) is a highly toxic waste product of pesticide manufacturing. It is resistant to biodegradation and highly sorptive to sediment. Sediments at the bottom of the reservoir behind L & D #2 in the Upper Mississippi River have been found to contain high C_6Cl_6 concentrations that indicate a large spill in the past. The sedimentation rate is largely unknown; however, an effective diffusion coefficient using a nonsorptive tracer and a retardation coefficient can be determined for C_6Cl_6 from core samples of the reservoir sediments. For hexachlorobenzene and these sediments, $D = 1.2 \times 10^{-9} \ \mathrm{m^2/s}$ and $R = 8{,}000$. As part of a forensic engineering exercise, and given the data, when did the spill occur?

Measured data for hexachlorobenzene spill

Depth (cm)	0	1	2	3	4	5	6	7
Conc. (g/m^3)	0	0	0	0.01	0.04	0.33	0.82	1.43
Depth (cm)	8	9	10	11	12	13	14	15
Conc. (g/m^3)	2.27	2.5	2.12	1.64	0.93	0.21	0.05	0.01

5. A barrel of 10 kg of methylchloride (CH_3Cl, a priority pollutant) was deposited many years ago at the bottom of a lake. Now it is 10 cm below the sediment surface and has begun to leak. What will be the concentration versus time and distance solution for this problem? Plot the concentration profile after 20 years. How would you determine the flux of methylchloride into the lake?

   Assume:

   $K_{ow} = 8, \rho_b/\varepsilon = 6 \, g/cm^3, f = 0.05$
   No flow in the sediments below the lake
   Diffusivity of CH_3Cl in sediments $= 10^{-9}/m^2s$

6. Ethyl chloroacetate, $C_4H_7ClO_2$, is used as a solvent and in the synthesis of intermediate dye chemicals. The effluent from a dye synthesis plant is discharged into a wetland, with a pH of 7.3. At this pH, the alkaline hydrolysis rate constant is 1.56 $M^{-1}s^{-1}$. If the alkalinity of the wetland is close to constant, compared with the concentration change of the ethyl chloroacetate, at 0.002 M, what would be the first-order rate constant for ethyl chloroacetate degradation?

7. In groundwater and soil pollution problems, there is sometimes discussion of fast sorption and slow sorption, where the local equilibrium assumption would not be valid. How would you formulate a diffusion equation to deal with both the fast and slow forms of adsorption and desorption?

3 Diffusion Coefficients

This chapter will discuss the determination of diffusion coefficients in dilute solutions (i.e., the concentration of the solute is dilute). The most important factor in diffusion coefficients is the type of media that the solute is diffusing through: gas, liquid, or solid. Some order-of-magnitude values of diffusion coefficients are presented in Table 3.1, which indicates that the diffusion coefficient of a compound through a gas is four orders of magnitude greater than through a liquid and nine orders of magnitude greater than through a solid. This can make a significant difference in the rate of diffusion in the different media classifications. In fact, the manner in which the important physical process used as a model to describe "transport" is quite different for gases, liquids, and solids.

A. Diffusion Coefficients in Gases

Diffusion coefficients of a compound through a gas will be described with the simple *kinetic theory of diffusion in gases*. We could call this the "pinball theory of diffusion." Although it can be a frustrating game to play, one can enjoy the mechanical aspects of the pinball bouncing around on the table. When we think of diffusion in gases, we can simply think of many pinballs bouncing around simultaneously after hitting the right trigger in the game. The description below may leave you with similar analogies.

Let us assume a linear concentration gradient over a distance Δx, with the change in concentration over this distance equal to ΔC, as shown in Figure 3.1. If we assume perfectly inelastic collisions (a superball approaches an inelastic collision by bouncing almost back to its drop height), and assign the molecules moving in the negative x-direction v_x^- and the molecules moving in the positive x-direction v_x^+, then we can say that the flux, J, is given by

$$J = \left[-1/2 \left(C_0 + \Delta C \right) v_x^- + 1/2 \, C_0 \, v_x^+ \right] L/\Delta x + \text{secondary effects} \qquad (3.1)$$

where v_x^- is the velocity of molecules moving in the $-x$- direction, v_x^+ is the velocity of molecules moving in the $+x$-direction, L is the mean free path of the molecules (or the mean distance before a collision with another molecule), and secondary effects

Table 3.1: *Order of magnitude values of diffusion coefficients*

Media	D		
Gases	10^{-5} m^2/s	or	0.1 cm^2/s
Liquids	10^{-9} m^2/s	or	10^{-5} cm^2/s
Plastics, glass	10^{-12} m^2/s	or	10^{-8} cm^2/s
Crystalline solids	10^{-14} m^2/s	or	10^{-10} cm^2/s

refer to the occurrence of two collisions within the distance Δx. In equation (3.1), we have split the molecules moving in the positive and negative x-directions evenly, such that exactly half the molecules at a certain location are moving in each direction.

We will now justify the term $L/\Delta x$ in equation (3.1). Let us assume for this purpose that Δx is small, such that $L/\Delta x = 2$, as demonstrated in Figure 3.2. We will be following four equally spaced molecules, moving at equal velocity magnitude, with two moving in the negative x-direction. The flux through the walls of the control volume is four molecules through each wall. Now, if we expand the control volume by a factor of two, leaving everything else identical as shown in Figure 3.2, the flux through each wall is reduced by a factor of two. Thus, the term $L/\Delta x$ must be in the equation for net flux through our control volume.

Now, it is a good assumption that the mean velocity of the molecules in the positive and negative directions is equal, or

$$v_x^- = v_x^+ \tag{3.2}$$

and equation (3.1) becomes

$$J = -1/2 \frac{\Delta C}{\Delta x} v_x L + \text{secondary effects} \tag{3.3}$$

We will be estimating the velocity of a molecule from kinetic theory, not the x-component of this velocity. Thus, from the Pythagoras theorem, the velocity of a molecule, v, is given by

$$v = \left(v_x^2 + v_y^2 + v_z^2\right)^{1/2} \tag{3.4}$$

and $v_x = v_y = v_z$ is a good assumption. Then, equation (3.4) becomes

$$v = \sqrt{3}\, v_x \tag{3.5}$$

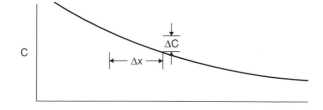

C

Figure 3.1. Linear concentration gradient over Δx.

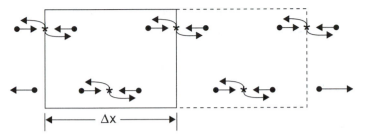

Figure 3.2. Illustration of the term $L/\Delta x$ in equation (3.1).

and equation (3.1) becomes

$$J = \frac{-1}{2\sqrt{3}} \frac{\Delta C}{\Delta x} vL + \text{secondary effects} \qquad (3.6)$$

If we calculate the secondary effects from statistical mechanics, and make Δx small, equation (3.6) becomes

$$J = -\frac{1}{3} \frac{\partial C}{\partial x} vL \qquad (3.7)$$

We have previously developed another equation for J, as well, that we will use here:

$$J = -D\frac{\partial C}{\partial x} \qquad (2.2)$$

Then, setting the right-hand side of the right-hand side of equation (3.7) equal to the left-hand side of equation (2.2), gives

$$D = vL/3 \qquad (3.8)$$

The kinetic theory of gases has therefore given us a fairly simple equation for the diffusion coefficient of a molecule. All that remains to be determined is the mean velocity of the molecule and the mean free path of the molecule.

The *mean free path* of a molecule has been computed (Kauzmann, 1966) to be

$$L = \frac{K_B T}{\sqrt{2}\,\pi\, d_{12}^2\, P} \qquad (3.9)$$

where d_{12} is the arithmetic mean diameter of the solute and solvent molecules, respectively, or $d_{12} = 1/2\,(d_1 + d_2)$; d_1 and d_2 are the equivalent spherical diameters of the solute and solvent molecules, respectively; P is the pressure of the gas; T is the absolute temperature; and K_B is Boltzman's constant ($K_B = 1.38 \times 10^{-23}$ m^2 kg/(s^2 °K)), which relates temperature and the kinetic energy of the molecule. Molecule diameters are given as σ in Table 3.2, where 1 angstrom $= 10^{-10}$ m.

Table 3.2: *Lennard-Jones potential parameters found from viscosities (Cussler, 1997)*

Substance		σ (Å)	ε/k (°K)
Ar	Argon	3.542	93.3
He	Helium	2.551	10.22
Kr	Krypton	3.655	178.9
Ne	Neon	2.820	32.8
Xe	Xenon	4.047	231.0
Air	Air	3.711	78.6
Br_2	Bromine	4.296	507.9
CCl_4	Carbon tetrachloride	5.947	322.7
CF_4	Carbon tetrafluoride	4.662	134.0
$CHCl_3$	Chloroform	5.389	340.2
CH_2Cl_2	Methylene chloride	4.898	356.3
CH_3Br	Methyl bromide	4.118	449.2
CH_3Cl	Methyl chloride	4.182	350
CH_3OH	Methanol	3.626	481.8
CH_4	Methane	3.758	148.6
CO	Carbon monoxide	3.690	91.7
CO_2	Carbon dioxide	3.941	195.2
CS_2	Carbon disulfide	4.483	467
C_2H_2	Acetylene	4.033	231.8
C_2H_4	Ethylene	4.163	224.7
C_2H_6	Ethane	4.443	215.7
C_2H_3Cl	Ethyl chloride	4.898	300
C_2H_5OH	Ethanol	4.530	362.6
CH_3OCH_3	Methyl ether	4.307	395.0
CH_3CHCH_2	Propylene	4.678	298.9
CH_3CCH	Methyl acetylene	4.761	251.8
C_3H_6	Cyclopropane	4.807	248.9
C_3H_8	Propane	5.118	237.1
$n\text{-}C_3H_2OH$	n-Propyl alcohol	4.549	576.7
CH_3COCH_3	Acetone	4.600	560.2
CH_3COOCH_3	Methyl acetate	4.936	469.8
$n\text{-}C_4H_{10}$	n-Butane	4.687	531.4
$iso\text{-}C_4H_10$	Isobutane	5.278	330.1
$C_2H_5OC_2H_5$	Ethyl ether	5.678	313.8
$CH_3COOC_2H_5$	Ethyl acetate	5.205	521.3
$n\text{-}C_5H_{12}$	n-Pentane	5.784	341.1
$C(CH_3)_4$	2,2-Dimethylpropane	6.464	193.4
C_6H_6	Benzene	5.349	412.3
C_6H_{12}	Cyclohexane	6.182	297.1
$n\text{-}C_6H_{14}$	n-Hexane	5.949	399.3
Cl_2	Chlorine	4.217	316.0
F_2	Fluorine	3.357	112.6
HBr	Hydrogen bromide	3.353	449
HCN	Hydrogen cyanide	3.630	569.1
HCl	Hydrogen chloride	3.339	344.7
HF	Hydrogen fluoride	3.148	330
HI	Hydrogen iodide	4.211	288.7
H_2	Hydrogen	2.827	59.7
H_2O	Water	2.641	809.1
H_2O_2	Hydrogen peroxide	4.196	289.3

Substance		σ (Å)	$\varepsilon/k(^\circ K)$
H_2S	Hydrogen sulfide	3.623	301.1
Hg	Mercury	2.969	750
I_2	Iodine	5.160	474.2
NH_3	Ammonia	2.900	558.3
NO	Nitric oxide	3.492	116.7
N_2	Nitrogen	3.798	71.4
N_2O	Nitrous oxide	3.828	232.4
O_2	Oxygen	3.467	106.7
PH_3	Phosphine	3.981	251.5
SO_2	Sulfur dioxide	4.112	335.4
UF_6	Uranium hexafluoride	5.969	236.8

Source: Data from Hirschfelder et al. (1954).

The mean velocity of a molecule can be similarly estimated from kinetic theory (Kauzmann, 1966):

$$v = \sqrt{\frac{8\,K_B T\,N_A}{\pi\,M_1}} \tag{3.10}$$

where M_1 is the molecular weight of the diffusing molecule and N_A is Avogadro's number, or the number of molecules per mole ($N_A = 6.02 \times 10^{23}$ molecules/mole).

EXAMPLE 3.1: *Mean free path and mean velocity of a diffusing gas*

The highly toxic pesticide, methyl bromide or CH_3Br, is diffusing through air. Estimate the mean velocity and mean free path of the methyl bromide molecules at an atmospheric pressure of 1 atm and a room temperature of 20°C.

$$d_1 = 4.118 \text{ Å from Table 3.2}$$
$$d_2 = 3.711 \text{ Å from Table 3.2}$$
$$d_{12} = 1/2(d_1 + d_2) = 3.92 \text{ Å or } 3.92 \times 10^{-10} \text{ m}$$
$$P = 1\,\text{atm} = 1.013 \times 10^3 \text{ N/m}^2 \text{ or kg/m-s}^2$$
$$T = 273 + 20 = 293^\circ\text{K}$$
$$M_1 = M(CH_3Br) = 12 + 3 + 37 = 52 \text{ g/mole}$$

$$L = \frac{(1.38 \times 10^{-23} \text{ kg m}^2/\text{s} - {}^\circ\text{K})(293^\circ\text{K})}{\sqrt{2}\,\pi\,(3.92 \times 10^{-10} \text{ m})^2\,(1.013 \times 10^3 \text{ kg/m-s}^2)} = 5.91 \times 10^{-6} \text{ m}$$

$$v = \sqrt{\frac{8\,(1.38 \times 10^{-23} \text{ kg m}^2/\text{s}^2 - {}^\circ\text{K})(293^\circ\text{K})(6.02 \times 10^{23}/\text{mole})}{\pi\,(52 \text{ g/mole})(1{,}000 \text{ kg/g})}}$$

$$= 0.351 \text{ m/s}$$

Although the mean velocity of the molecules is substantial, the typical distance between collisions is rather small ($\sim$0.6 μm). Each molecule would thus experience roughly $v/L = 6 \times 10^5$ collisions per second.

Finally, substitution of equations (3.9) and (3.10) into equation (3.8) results in an equation for the diffusivity of a gas diffusing through another gas:

$$D_{12} = \frac{2}{3} \left(\frac{K_B}{\pi} \right)^{3/2} N_A^{1/2} \frac{T^{3/2}}{P \, d_{12}^2 M_1^{1/2}} \tag{3.11}$$

where D_{12} is the diffusion coefficient for the diffusion of gas 1 through gas 2.

Although the kinetic theory of gases is not generally used, as is, to estimate the diffusion coefficients of gases, it does provide a framework for the characterization of data through predictive equations. This simple kinetic theory has shown us the following:

1. $D \sim 1/P$. Higher pressure generally is due to more molecules per volume, so the number of molecules is proportional to pressure, and diffusivity is inversely proportional to the number of molecules. More molecules means less distance before the diffusing molecule hits another molecule and bounces back.

2. $D \sim 1/d_{12}^2$. The projected area of a molecule is proportional to d^2. Diffusivity is shown to be inversely proportional to the mean projected area of the diffusing and ambient molecules.

3. $D \sim 1/M_1^{1/2}$. The inertia of a large molecule will be greater than a small molecule; so, a large molecule will require more energy to reach a given velocity. Equation (3.11) indicates that this inertial effect is equal to $M_1^{1/2}$.

4. $D \sim T^{3/2}$. Equations (3.9) and (3.10) indicate that both the velocity and mean free path of the molecule are dependent on the kinetic energy of the molecules. The mean free path is affected because molecules that are moving faster will usually travel farther, as well.

1. Chapman–Enskog Equation for Diffusion of Gases

The Chapman–Enskog equation (see Chapman and Cowling, 1970) is semi-empirical because it uses equation (3.11) and adjusts it for errors in the observations of diffusivity in gases. It also includes a parameter, Ω, to account for the elasticity of molecular collisions:

$$D_{12}(\text{cm}^2/\text{s}) = \frac{\beta \, T^{3/2} \left(\dfrac{1}{M_1} + \dfrac{1}{M_2} \right)^{1/2}}{P \, d_{12}^2 \, \Omega} \tag{3.12}$$

where Ω is a collision integral that must be found from look-up tables, such as that given in Tables 3.2 and 3.3, M_2 is the molecular weight of the gas that compound 1 is diffusing through, and β is a constant value of 1.83×10^{-3}. We must be careful about the units in equation (3.12), because the constant, β, has units. With $\beta = 1.8 \times 10^{-3}$, d_{12} is given in angstroms, P in atmospheres, and T in $^\circ$K. Equation (3.12) is accurate to within roughly 8%, as long as the molecule is nonpolar. Equation (3.12) would therefore be less accurate for estimating the diffusivities of compounds like water vapor, carbon monoxide, and ammonia, because these are highly polar molecules.

Table 3.3: *The collision integral Ω (Cussler, 1997)*

kT/ε	Ω	kT/ε	Ω	kT/ε	Ω
0.30	2.662	1.65	1.153	4.0	0.8836
0.35	2.476	1.70	1.140	4.1	0.8788
0.40	2.318	1.75	1.128	4.2	0.8740
0.45	2.184	1.80	1.116	4.3	0.8494
0.50	2.066	1.85	1.105	4.4	0.8652
0.55	1.966	1.90	1.094	4.5	0.8610
0.60	1.877	1.95	1.084	4.6	0.8568
0.65	1.798	2.00	1.075	4.7	0.8530
0.70	1.729	2.1	1.057	4.8	0.8492
0.75	1.667	2.2	1.041	4.9	0.8456
0.80	1.612	2.3	1.026	5.0	0.8422
0.85	1.562	2.4	1.012	6	0.8124
0.90	1.517	2.5	0.9996	7	0.7896
0.95	1.476	2.6	0.9878	8	0.7712
1.00	1.439	2.7	0.9770	9	0.7556
1.05	1.406	2.8	0.9672	10	0.7424
1.10	1.375	2.9	0.9576	20	0.6640
1.15	1.346	3.0	0.9490	30	0.6232
1.20	1.320	3.1	0.9406	40	0.5960
1.25	1.296	3.2	0.9328	50	0.5756
1.30	1.273	3.3	0.9256	60	0.5596
1.35	1.253	3.4	0.9186	70	0.5464
1.40	1.233	3.5	0.9120	80	0.5352
1.45	1.215	3.6	0.9058	90	0.5256
1.50	1.198	3.7	0.8998	100	0.5130
1.55	1.182	3.8	0.8942	200	0.4644
1.60	1.167	3.9	0.8888	300	0.4360

Source: Data from Hirschfelder et al. (1954).

2. Wilke–Lee Adjustment to the Chapman–Enskog Equation

Wilke and Lee (1955) found that the Chapman–Enskog equation could not estimate the diffusivity of lower molecular weight compounds, as well as those with a higher molecular weight. They therefore adjusted the constant, β, as follows:

$$\beta = 2.17 \times 10^{-3} - 5 \times 10^{-4} \left(\frac{1}{M_1} + \frac{1}{M_2} \right)^{1/2} \tag{3.13}$$

where β has the same messy units as in equation (3.12).

EXAMPLE 3.2: *Influence of temperature on diffusion coefficient*

How does the diffusivity of water vapor in air at 5°C compare with that at 25°C?

All three equations discussed in this text provide the following dependence of diffusivity on temperature:

$$D \sim T^{3/2}$$

Thus, we can relate the ratio of diffusivities as follows:

$$\frac{D_5}{D_{20}} = \left(\frac{273+5}{273+20}\right)^{3/2} = 0.9$$

EXAMPLE 3.3: *Estimating gaseous diffusion coefficients*

Estimate the diffusivity of water vapor in air at 25°C and 1 atm. Compare these estimates with measured values, taken from Table 3.4.

Chapman–Enskog equation:

$$D_{12}(\text{cm}^2/\text{s}) = 1.83 \times 10^{-3} \frac{T^{3/2}\left(\dfrac{1}{M_1} + \dfrac{1}{M_2}\right)^{1/2}}{P\,d_{12}^2\,\Omega} \tag{3.12}$$

where subscript 1 = water (H_2O), and subscript 2 = air. Then, $T = 298°K$, $M_1 = 18$ g/mole, $M_2 = 29$ g/mole, and $P = 1$ atm. From Table 3.2:

$$d_1 = 2.64 \text{ Å}$$
$$d_2 \approx 0.8(3.8) + 0.2(3.5) = 3.74 \text{ Å}$$
$$d_{12} = 1/2(d_1 + d_2) = 3.17 \text{ Å}$$
$$\varepsilon_1/k = 809°K$$
$$\varepsilon_2/k \approx 0.8(71.4) + 0.2(106.7) = 78.5°K$$
$$\varepsilon_{12}/k = \sqrt{\frac{\varepsilon_1}{k}\frac{\varepsilon_2}{k}} = 252°K$$
$$\therefore kT/\varepsilon = 298°K/252°K = 1.18$$

From Table 3.3:

$$\Omega \cong 1.33$$

and equation (3.12) gives

$$D_{12} = 0.21 \text{ cm}^2/\text{s} = 2 \times 10^{-5} \text{ m}^2/\text{s}$$

(Chapman–Enskog equation)

Now try the Wilke–Lee adjustment:
The constant in equation (3.12) is replaced by

$$\beta = 2.17 \times 10^{-3} - 5 \times 10^{-4}\left(\frac{1}{M_1} + \frac{1}{M_2}\right)^{1/2} \tag{3.13}$$

or

$$\beta = 2.0 \times 10^{-3} \text{ for } H_2O \text{ diffusing in air.}$$

Table 3.4: *Experimental values of diffusion coefficients in gases at 1 atm (Cussler, 1997)*

Gas pair	Temperature (°K)	Diffusion coefficient $(cm^2\,s^{-1})$
Air–CH_4	273.0	0.196
Air–C_2H_5OH	273.0	0.102
Air–CO_2	276.2	0.142
	317.2	0.177
Air–H_2	273.0	0.611
Air–D_2	296.8	0.565
Air–H_2O	289.1	0.282
	298.2	0.260
	312.6	0.277
	333.2	0.3050
Air–He	276.2	0.6242
Air–O_2	273.0	0.1775
Air–n-hexane	294	0.080
Air–n-heptane	294	0.071
Air–bezene	298.2	0.096
Air–toluene	299.1	0.0860
Air–chlorobenzene	299.1	0.074
Air–aniline	299.1	0.074
Air–nitrobenzene	298.2	0.0855
Air–2-propanol	299.1	0.099
Air–butanol	299.1	0.087
Air–2-butanol	299.1	0.089
Air–2-pentanol	299.1	0.071
Air–ethylacetate	299.1	0.087
CH_4–Ar	298	0.202
CH_4–He	298	0.675
CH_4–H_2	298.0	0.726
CH_4–H_2O	307.7	0.292
CO–N_2	295.8	0.212
^{12}CO–^{14}CO	373	0.323
CO–H_2	295.6	0.7430
CO–D_2	295.7	0.5490
CO–He	295.6	0.7020
CO–Ar	295.7	0.1880
CO_2–H_2	298.0	0.6460
CO_2–N_2	298.2	0.165
CO_2–O_2	293.2	0.160
CO_2–He	298	0.612
CO_2–Ar	276.2	0.1326
CO_2–CO	296.1	0.1520
CO_2–H_2O	307.5	0.202
CO_2–N_2O	298.0	0.117
CO_2–SO_2	263	0.064
$^{12}CO_2$–$^{14}CO_2$	312.8	0.125
CO_2–propane	298.0	0.0863
CO_2–ethyleneoxide	298.0	0.0914

(continued)

Table 3.4 *(continued)*

Gas pair	Temperature ($^\circ$K)	Diffusion coefficient ($cm^2 s^{-1}$)
H_2–N_2	297.2	0.779
H_2–O_3	273.2	0.697
H_2–D_2	288.2	1.24
H_2–He	298.2	1.132
H_2–Ar	287.9	0.828
H_2–Xe	341.2	0.751
H_2–SO_2	285.5	0.525
H_2–H_2O	307.1	0.915
H_2–NH_3	298	0.783
H_2–acetone	296	0.424
H_2–ethane	298.0	0.537
H_2–*n*-butane	287.9	0.361
H_2–*n*-hexane	288.7	0.290
H_2–cyclohexane	288.6	0.319
H_2–benzene	311.3	0.404
H_2–SF_4	286.2	0.396
H_2–*n*-heptane	303.2	0.283
H_2–*n*-decane	364.1	0.306
N_2–O_2	273.2	0.181
	293.2	0.22
N_2–He	298	0.687
N_2–Ar	293	0.194
N_2–NH_3	298	0.230
N_2–H_2O	307.5	0.256
N_2–SO_2	263	0.104
N_2–ethylene	298.0	0.163
N_2–ethane	298	0.148
N_2–*n*-butane	298	0.096
N_2–isobutane	298	0.0905
N_2–*n*-hexane	288.6	0.076
N_2–*n*-octane	303.1	0.073
N_2–2,2,4-trimethylpentane	303.3	0.071
N_2–*n*-decane	363.6	0.084
N_2–benzene	311.3	0.102
O_2–He (He trace)	298.2	0.737
(O_2 trace)	298.2	0.718
O_2–He	298	0.729
O_2–H_2O	308.1	0.282
O_2–CCl_4	296	0.075
O_2–benzene	311.3	0.101
O_2–cyclohexane	288.6	0.075
O_2–*n*-hexane	288.6	0.075
O_2–*n*-octane	303.1	0.071
O_2–2,2,4-trimethylpentane	303.0	0.071
He–D_2	295.1	1.250
He–Ar	298	0.742
He–H_2O	298.2	0.908
He–NH_3	297.1	0.842

Gas pair	Temperature (°K)	Diffusion coefficient ($cm^2 s^{-1}$)
He–n-hexane	417.0	0.1574
He–benzene	298.2	0.384
He–Ne	341.2	1.405
He–methanol	423.2	1.032
He–ethanol	298.2	0.494
He–propanol	423.2	0.676
He–hexanol	423.2	0.469
Ar–Ne	303	0.327
Ar–Kr	303	0.140
Ar–Xe	329.9	0.137
Ar–NH_3	295.1	0.232
Ar–SO_2	263	0.077
Ar–n-hexane	288.6	0.066
Ne–Kr	273.0	0.223
Ethylene–H_2O	307.8	0.204
Ethane–n-hexane	294	0.0375
N_2O–propane	298	0.0860
N_2O–ethyleneoxide	298	0.0914
NH_3–SF_6	296.6	0.1090
Freon-12–H_2O	298.2	0.1050
Freon-12–benzene	298.2	0.0385
Freon-12–ethanol	298.2	0.0475

Source: Data from Hirschfelder et al. (1954) and Reid et al. (1977).

Substituting the constant from equation (3.13) into equation (3.12) gives

$$D_{12} = 0.23 \text{ cm}^2/\text{s} = 2.3 \times 10^{-5} \text{ m}^2/\text{s}$$
(Wilke–Lee adjustment)

Table 3.4 gives a measured value for D_{12}:

$$D_{12} \text{ (Table 3.4)} = 0.26 \text{ cm}^2/\text{s} = 2.6 \times 10^{-5} \text{ m}^2/s$$

We can now compare the predictions to the measurements:

Formula for D_{12}	Measured-predicted/ measured
Chapman–Enskog	0.19
Wilke–Lee	0.12

The Chapman–Enskog equation missed the measurement by 19% and the Wilke–Lee adjustment by 12%. Both of these are greater than the 8% mentioned previously, but water is a highly polar molecule.

B. Diffusion Coefficients in Liquids

Liquids are denser than gases (roughly three orders of magnitude at atmospheric pressure), so the mean free path concept developed from the kinetic theory of gases does not work well for liquids. The mean free path is virtually zero, because the molecules are always bumping into each other. There are two analogies that describe this well. The first involves rugby. Diffusion in gases is similar to the activities outside the scrum, where players can run a certain distance before they are tackled, pass the ball, or kick the ball. This corresponds to a free path, and speed is an important component. Diffusion through liquids, however, is more like being in the middle of a scrum. Each player is in continuous contact with other players in the scrum, and they do not have any free path to speak of. The second analogy involves American football. Diffusion through gases is analogous to being a wide receiver, downfield. Although there is contact, it is typically one defender against one wide receiver, similar to the kinetic theory of gases, although the collisions are often far from inelastic. Diffusion through liquids is analogous to being a lineman. There is constant bumping, and sometimes pushing and holding, as the offensive and defensive lines try to gain some advantage. In a liquid, one molecule is never far from another, and the contact between molecules is almost continuous, relative to the molecule contact in gases.

1. Einstein–Stokes Relationship

Since diffusion in liquids involves one molecule in close contact with other molecules, Einstein (1905) developed the following relationship:

$$D_{12} = K_B T / \xi_{12} \tag{3.14}$$

where ξ_{12} is a friction coefficient that represents the friction impeding the movement of the solute (1) through the solvent (2). Because these molecules are moving slowly, Stokes' law of frictional drag (Stokes, 1851) applies, and

$$\xi_{12} = 3\pi \, \mu_2 d_1 \tag{3.15}$$

where μ_2 is the viscosity of the solvent and d_1 is the equivalent spherical diameter of the solute molecule. Then, substituting equation (3.15) into (3.14),

$$D_{12} = K_B T / (3\pi \mu_2 d_1) \tag{3.16}$$

Equation (3.16) is generally accurate to within 20%, with some notable exceptions, and is one of the standards by which other equations are compared. Note that equation (3.16) also provides a characterization of D_{12} with the following parameters:

$$D_{12} \sim T$$

Because absolute temperature does not change much for environmental applications, this is the least variable characterization of equation (3.16).

$$D_{12} \sim 1/\mu_2$$

Highly viscous solvents, like molasses or glycerin, will have a lower diffusivity than solvents like kerosene or alcohol, all other factors being equal. This is also where temperature becomes important, because the viscosity of a liquid is highly dependent on temperature.

$$D_{12} \sim 1/d_1$$

In general, large molecules (with a larger equivalent diameter) will have a lower diffusion coefficient than small molecules. Thus, the diffusion coefficient for hydrogen and helium are approximately an order of magnitude greater than for the long-chained organic compounds.

2. Wilke–Chang Relationship

Wilke and Chang (1955) developed an empirical relationship that was based on the temperature and viscosity characterization of the Stokes–Einstein relationship. It deviates from the equivalent diameter characterization by using another parameter, and incorporates the size of the solvent molecule and a parameter for polarized solvents. It is the most generally used of the available equations (Lyman et al., 1990) and is given as

$$D_{12}(\text{cm}^2/\text{s}) = 7.4 \times 10^{-8} \frac{T}{\mu_2} \frac{(X M_2)^{1/2}}{V_b^{0.6}} \tag{3.17}$$

where V_b is the molar volume (cm^3/mole) of the solute at the boiling point, M_2 is the molecular weight of the solvent, and the units for μ_2 and T are centipoise (or g/cm-s $\times 10^{-2}$) and °K, respectively. Finally, X is an association parameter for the solvent polarity. Some values of X are given in Table 3.5.

If the molar volume is not known, one of two relationships can be used:

$$V_b = 0.285 \, V_c^{1.048} \tag{3.18}$$

Table 3.5: *Examples of the association parameter in the Wilke–Chang equation for diffusion in liquids (Wilke and Chang, 1955)*

X	Solvent	
2.6	Water	Polar
2.26	Water (Hayduk and Laudie, 1974)	
1.9	Methanol	⇓
1.5	Ethanol	
1.0	Benzene, ether, heptone	Nonpolar

Table 3.6: *Additive volume increments for calculating LeBas molar volume, V_b' (Cussler, 1997)*

Atom	Increment (cm³/mole).	Atom	Increment (cm³/mole).
C	14.8	Br	27.0
H	3.7	Cl	24.6
O (except as noted below)	7.4	F	8.7
In methyl esters and ethers	9.1	I	37.0
In ethyl esters and ethers	9.9	S	25.6
In higher esters and ethers	11.0	Ring	
In acids	12.0	Three-membered	−6.0
Joined to S, P, N	8.3	Four-membered	−8.5
N		Five-membered	−11.5
Double bonded	15.6	Six-membered	−15.0
In primary amines	10.5	Naphthalene	−30.0
In secondary amines	12.0	Athracene	−47.5

Source: Le Bas, 1915.

or

$$V_b = 0.9 \, V_b' \tag{3.19}$$

where V_c is the critical volume (cm³/mole) or V_b at critical temperature and pressure, and V_b' is the "LeBas" molar volume (cm³/mole)(LeBas, 1915). The critical volume is given for a number of compounds by Reid et al. (1977). Computation of the LeBas molar volume, described below, uses the additive volume increments given in Table 3.6.

EXAMPLE 3.4: *LeBas molar volume*

Calculate the LeBas molar volume for toluene (methyl benzene).

Determine the structural formula and the crystal structure for the compound:

$C_6H_5CH_3 \Rightarrow$ Six-membered ring with one methyl group

Use Table 3.5 to compute the additive increments:

C_7	=	7(14.8)	=	103.6
H_8	=	8(3.7)	=	29.6
Six-membered ring			=	−15.0

V_b	=	118.2 cm³/mole

3. Hayduk–Laudie Relationship for Diffusion Coefficient in Water

Hayduk and Laudie (1974) developed a relationship specifically for water. They eliminated some of the solvent-specific parameters from the Wilkie–Chang relationship, eliminated absolute temperature, and fit three coefficients in the relationship. The results were as follows:

$$D_{12}(\text{cm/s}) = \frac{1.326 \times 10^{-4}}{\mu_2^{1.14} V_{b_1}'^{\,0.589}} \tag{3.20}$$

where μ_2 is in centipoise and V_{b_1}' is the LeBas molar volume (cm^3/mole) of the solute at the boiling point.

EXAMPLE 3.5: *Estimating diffusion coefficient in liquids*

Estimate the diffusivity of toluene in water at 25°C using the Wilke–Chang and the Hayduk–Laudie relationships. Compare your result with measured values in Table 3.7.

Wilke–Chang relationship:

$$D_{12}(\text{cm}^2/\text{s}) = 7.4 \times 10^{-8} \frac{T}{\mu_2} \frac{(X M_2)^{1/2}}{V_b^{0.6}} \tag{3.17}$$

$$T = 298°\text{K}$$
$$X = 2.26 \text{ for water (Hayduk–Laudie)}$$
$$M_2 = 18 \text{ g/mole}$$
$$\mu_2 = 0.894 \text{ cp}$$
$$V_{b1} = 0.285 \, V_c^{1.048}$$

Lyman et al. (1990) gives $V_c = 316$ for toluene
$\quad \therefore \; V_b = 118 \text{ cm}^3$/mole
$\quad$ (*Note*: We estimated $V_{b_1}' = 118.2 \text{ cm}^3$/mole in Example 3.4.)

$$V_b = 0.9 \, V_b' \Rightarrow 106 \text{ cm}^3/\text{mole}$$
$$\Rightarrow 6\% \text{ difference in } D_{12}$$

Thus, equation (3.17) gives

$$D_{12} = 0.90 \times 10^{-5} \text{ cm}^2/\text{s} = 0.9 \times 10^{-9} \text{ m}^2/\text{s}$$

Hayduk–Laudie relationship:

$$D_{12} \left(\text{cm}^2/\text{s}\right) = \frac{1.326 \times 10^{-4}}{\mu_2^{1.14} \left(V_b'\right)^{0.589}} \tag{3.20}$$

Table 3.7A: *Diffusion coefficients at infinite dilution in water at 25°C (Cussler, 1997)*

Solute	D ($\times 10^{-5}$ cm^2/s)
Argon	2.00
Air	2.00
Bromine	1.18
Carbon dioxide	1.92
Carbon monoxide	2.03
Chlorine	1.25
Ethane	1.20
Ethylene	1.87
Helium	6.28
Hydrogen	4.50
Methane	1.49
Nitric oxide	2.60
Nitrogen	1.88
Oxygen	2.10
Propane	0.97
Ammonia	1.64
Benzene	1.02
Toluene	0.95
Hydrogen sulfide	1.41
Sulfuric acid	1.73
Nitric acid	2.60
Acetylene	0.88
Methanol	0.84
Ethanol	0.84
1-Propanol	0.87
2-Propanol	0.87
n-Butanol	0.77
Benzyl alcohol	0.821
Formic acid	1.50
Acetic acid	1.21
Propionic acid	1.06
Benzoic acid	1.00
Glycine	1.06
Valine	0.83
Acetone	1.16
Urea	$(1.380 - 0.0782c_1, + 0.00464c_1^2)^*$
Sucrose	$(0.5228 - 0.265C)^*$
Ovalbumin	0.078
Hemoglobin	0.069
Urease	0.035
Fibrinogen	0.020

* Known to very high accuracy and often used for calibration; c_1 is the concentration of the solute in moles per liter.
Source: Data from Cussler (1976) and Sherwood et al. (1975).

Table 3.7B: *Diffusion coefficients at infinite dilation in nonaqueous liquids (Cussler, 1997)*

Solute*	Solvent	$D(\times 10^{-5}\ cm^2/s)$
Acetone	Chloroform	2.35
Benzene		2.89
n-Butyl acetate		1.71
Ethyl alcohol (15°)		2.20
Ethyl ether		2.14
Ethyl acetate		2.02
Methyl ethyl ketone		2.13
Acetic acid	Benzene	2.09
Aniline		1.96
Benzoic acid		1.38
Cyclohexane		2.09
Ethyl alcohol (15°)		2.25
n-Heptane		2.10
Methyl ethyl ketone (30°)		2.09
Oxygen (29.6°)		2.89
Toluene		1.85
Acetic acid	Acetone	3.31
Benzoic acid		2.62
Nitrobenzene (20°)		2.94
Water		4.56
Carbon tetrachloride	*n*-Hexane	3.70
Dodecane		2.73
n-Hexane		4.21
Methyl ethyl ketone (30°)		3.74
Propane		4.87
Toluene		4.21
Benzene	Ethyl alcohol	1.81
Camphor (20°)		0.70
Iodine		1.32
Iodobenzene (20°)		1.00
Oxygen (29.6°)		2.64
Water		1.24
Carbon tetrachloride		1.50
Benzene	*n*-Butyl alcohol	0.988
Biphenyl		0.627
p-Dichlorobenzene		0.817
Propane		1.57
Water		0.56
Acetone (20°)	Ethyl acetate	3.18
Methyl ethyl ketone (30°)		2.93
Nitrobenzene (20°)		2.25
Water		3.20
Benzene	*n*-Heptane	3.40

* Temperature 25°C except us indicated.
Source: Data from Rend et al. (1977).

We computed

$$V_b' = 118.2 \text{ cm}^3/\text{mole}$$

in Example 3.4. Thus, equation (3.20) gives

$$D_{12} = 0.91 \times 10^{-5} \text{ cm}^2/\text{s} = 0.91 \times 10^{-9} \text{ m}^2/\text{s}$$

Measured diffusivity of toluene in water at 25°C is given in Table 3.7:

$$D_{12} = 0.95 \times 10^{-5} \text{cm}^2/\text{s}$$

The predictions and measurements can be compared as

Formula	Measured-predicted/measured
Wilke–Chang	0.053
Hayduk–Laudie	0.042

Both equations seem to provide relative accuracy in this case.

C. Problems

1. Estimate the diffusion coefficient for 10 compounds through air at 1 atmosphere pressure from the Wilke–Lee adjustment to the Chapman–Enskog theory and compare your results with measurements. What is the percent error of the estimation (assuming that the measurements are correct)? What is the primary cause of the differences between the estimated diffusivities?

2. Estimate the diffusivities of the 10 compounds from problem 1 in air at Leadville, Colorado, in winter, when air temperature is −40°C and air pressure is 0.7 atm.

3. Estimate the diffusion coefficient of 10 compounds in water at 25°C with the Wilke–Chang theory using $X = 2.26$ and the Hayduk–Laudie theory, and compare with measured values. What is the overall percent error (mean of all absolute percent errors)?

4. Estimate the diffusivity of the 10 compounds from problem 3 at 1°C. Compare and explain the importance of temperature to diffusion in water versus the importance of temperature to diffusion in air.

5. Determine the ratio of diffusion coefficient at 5°C to diffusion coefficient at 20°C in water for methane, formic acid, and hydrogen sulfide.

4   Mass, Heat, and Momentum Transport Analogies

In Chapter 2, we used the control volume technique represented by equation (2.1) to transport mass into and out of our control volume. Inside of the control volume, there were source and sink rates that acted to increase or reduce the mass of the compound. Anything left after these flux and source/sink terms had to stay in the control volume, and was counted as accumulation of the compound.

$$\text{Flux rate} - \text{Flux rate} + \text{Source} - \text{Sink} = \text{Accumulation} \tag{2.1}$$
$$\text{IN} \qquad \text{OUT} \qquad \text{rate} \qquad \text{rate}$$

Equation (2.1) will also apply to the transport of any fluid property through our control volume, such as heat and momentum.

A. Heat Transport

The heat of a fluid per unit volume is given by

$$\text{Heat/Volume} = \rho C_p T \tag{4.1}$$

where C_p is the heat capacity of the fluid at constant pressure (cal/g/°C) and the diffusion of heat is described by a thermal diffusion coefficient, α:

$$\alpha = \frac{k_T}{\rho C_p} \tag{4.2}$$

where k_T is the thermal conductivity of heat (°C/cal/m/s). Applying a similar operation as we did for mass on the rectangular control volume, equation (2.14) becomes the heat transport equation:

$$\frac{\partial T}{\partial t} + \frac{\partial}{\partial x}(uT) + \frac{\partial}{\partial y}(vT) + \frac{\partial}{\partial z}(wT)$$
$$= \left[\frac{\partial}{\partial x}\left(\alpha \frac{\partial T}{\partial x}\right) + \frac{\partial}{\partial y}\left(\alpha \frac{\partial T}{\partial y}\right) + \frac{\partial}{\partial z}\left(\alpha \frac{\partial T}{\partial z}\right)\right] + \frac{S}{\rho C_p} \tag{4.3}$$

If the flow is incompressible, the resulting equation is analogous to equation (2.18):

$$\frac{\partial T}{\partial t} + u\frac{\partial T}{\partial x} + v\frac{\partial T}{\partial y} + w\frac{\partial T}{\partial z} = \alpha\left(\frac{\partial^2 T}{\partial x^2} + \frac{\partial^2 T}{\partial y^2} + \frac{\partial^2 T}{\partial z^2}\right) + \frac{S}{\rho C_p} \tag{4.4}$$

The most common sources and sinks of heat are short-wave radiation (sunshine), long-wave radiation (such as from a radiator), and heat sources and sinks from reactions. There are also boundary sources and sinks, such as evaporation and freezing. Application of equation (4.4) will be demonstrated through the following examples.

EXAMPLE 4.1: *Formation of ice on a lake surface (heat transfer with an abrupt change in boundary temperature)*

White Bear Lake in Minnesota is well mixed at 4°C before it freezes over. On the calm, cold night of December 1, the surface cools and a thin sheet of ice forms on the surface, followed by a 5-cm snow fall. This combination is sufficient to remain throughout the following days. An ice thickness of 20 cm is needed before you can bring your four-wheel-drive truck, with ice fishing house in tow, out on the ice to set up for a winter of ice fishing. Knowing that the average Minnesota temperature in December is −8°C, when can you expect to get that ice house out on the lake?

> *Ice properties*
> $C_p = 1.93$ W s/kg °K
> $\rho = 913$ kg/m^3
> $k_T = 2.22 \times 10^{-4}$ W/m °K

Therefore,

$$\alpha = \frac{k_T}{\rho C_p} = 1.24 \times 10^{-7}\, \text{m}^2/\text{s}$$

There is no flow in the ice, and the lake below the ice is assumed to be calm. The snow has precluded radiation from entering the ice, and mediates the high and low temperatures. In addition, the 5 cm of snow is equivalent to approximately 50 cm of ice, in terms of themal resistance. We will therefore set our depth of ice at 70 cm. Then, the following terms in equation (4.4) may be estimated:

1. $u = v = w = 0$
2. $\partial T/\partial x = \partial T/\partial y = 0$, where z is the vertical coordinate. This is because of exposure to similar boundary conditions at the ice surface.
3. $S = 0$

Then, equation (4.4) becomes

$$\frac{\partial T}{\partial t} = \alpha\frac{\partial^2 T}{\partial z^2} \tag{E4.1.1}$$

with boundary conditions:

1. At $t = 0$, $z \neq 0$, $T = 4°C$
2. At $t > 0$, $z = 0$, $T = -8°C$

The boundary conditions, while not exact, are close to those of Example 2.10. They would be truly analogous if we make the following transformation:

$$T^* = T - 4°C$$

Then,

1. At $t = 0$, $z \neq 0$, $T^* = 0°C$
2. At $t > 0$, $z = 0$, $T^* = T_0^* = -12°C$

Making the appropriate substitutions in equation (E2.10.5), the solution to equation (E4.1.1) is

$$T^* = T_0^* \left[1 - \mathrm{erf}\left(\frac{z}{\sqrt{4\alpha t}} \right) \right] = T_0^* \, \mathrm{erfc}\left(\frac{z}{\sqrt{4\alpha t}} \right) \tag{E4.1.2}$$

The water temperature will need to reach $-0.3°C$ to overcome the heat of fusion (84 cal/g). Thus, we will assume that the lower edge of the ice occurs at $T = -0.3°C$ or $T^* = -4.3°C$. Then, equation (E4.1.2) becomes

$$\frac{T^*}{T_0^*} = \frac{-4.3}{-12} = \mathrm{erfc}\left(\frac{z}{\sqrt{4\alpha t}} \right) \tag{E4.1.3}$$

or

$$\mathrm{erfc}\left(\frac{z}{\sqrt{4\alpha t}} \right) = 0.36 \tag{E4.1.4}$$

Appendix A–5 indicates that this value of the complimentary error function occurs at $\eta = 0.65$. Thus,

$$\frac{z}{\sqrt{4\alpha t}} = 0.65 \tag{E4.1.5}$$

or

$$t = \frac{(0.7\,\mathrm{m})^2}{4\,(0.65)^2 \, 1.24 \times 10^{-7}\,\mathrm{m^2/s}} = 2.3 \times 10^6\,\mathrm{s} \cong 27\,\mathrm{days} \tag{E4.1.6}$$

It would require roughly 27 days for sufficiently thick ice allowing us to pull the ice house onto the lake. Of course, it is not a bad idea to check with the newscasters on ice thickness, because the weather and snow cover varies substantially between years.

EXAMPLE 4.2: *Post hole depth to avoid frost heave (solution to the diffusion equation with oscillating boundary conditions)*

You are building a deck addition to your house in St. Paul, Minnesota, while your sister is doing the same in Boston, Massachusetts. In comparing notes on the designs, you discover that your post hole depth is significantly greater than hers. Why?

The primary reason to place a post at deeper depths in the northern midwest is frost heave. The post hole needs to be below the frost line. What is this depth in each city?

For this example, we will consider the soil surface as a boundary condition with an oscillating temperature, described by a cosine function. The soil will conduct heat from the surface, without flow. The only transport mechanism will be the thermal conduction of the soil matrix.

We will make the following assumptions:

1. There is no flow in the soil.
2. The uniform matrix of the ground is large, compared with the depth of interest; thus, $\partial^2 T / \partial x^2 \Rightarrow 0$ and $\partial^2 T / \partial y^2 \Rightarrow 0$ where z is depth.
3. There are no heat sources in the soil.
4. The time since initiation of the oscillating thermal boundary condition is large, so that initial conditions can be neglected.

Then, equation (4.3) or (4.4) becomes

$$\frac{\partial T}{\partial t} = \alpha \frac{\partial^2 T}{\partial z^2} \tag{E4.2.1}$$

similar to equation (E4.1.1), with boundary conditions:

1. At $z = 0$, $T = T_a + \Delta T \cos(nt)$
2. At $z = \infty$, $T = T_a$

where T_a is the average temperature of the surface and ΔT is the maximum deviation from the average.

The solution to equation (E4.2.1), with the boundary conditions listed above and assuming that initial conditions can be ignored, is as follows:

$$T = T_a + \Delta T e^{-\lambda z} \cos(nt - \lambda z) \tag{E4.2.2}$$

or

$$\theta = e^{-\lambda z} \cos(nt - \lambda z) \tag{E4.2.3}$$

where

$$\lambda = \sqrt{\frac{n}{2\alpha}} \tag{E4.2.4}$$

and

$$\theta = \frac{T - T_a}{\Delta T} \tag{E4.2.5}$$

Equation (E4.2.3) is plotted in Figure E4.2.1. This equation satisfies the governing equation when time is large and satisfies the boundary conditions. It is a damped

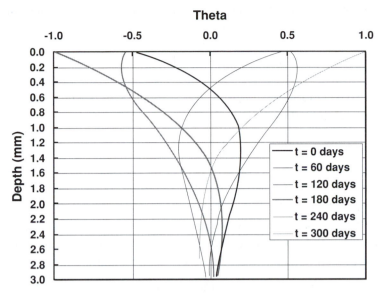

Figure E4.2.1. Illustration of solution to Example 4.4, with $\alpha = 10^{-7}\,\mathrm{m^2/s}$ and $n = 1.99 \times 10^{-7}\,\mathrm{s^{-1}}$.

progressive wave, where damping and the wave movement are in the same direction. In equation (E4.2.3), $e^{-\lambda z}$ is a damping factor and λz is a time lag.

We can further examine equation (E4.2.3) by determining the *wave peak*,

$$\frac{\partial T}{\partial z} = \Delta T \lambda e^{-\lambda z} \left[\sin(nt - \lambda z) - \cos(nt - \lambda z) \right] = 0 \qquad \text{(E4.2.6)}$$

or

$$\sin(nt - \lambda z) = \cos(nt - \lambda z) \qquad \text{(E4.2.7)}$$

which occurs at

$$nt - \lambda z = \pi/4 \ \text{(and } 5\pi/4) \qquad \text{(E4.2.8)}$$

which gives

$$z = \sqrt{\frac{2\alpha}{n}} \left(nt - \frac{\pi}{4} \right) \qquad \text{(E4.2.9)}$$

The *wave speed* is found from $\partial(\text{E4.2.9})/\partial t$ or

$$\text{wave speed} = \sqrt{2\alpha n} \qquad \text{(E4.2.10)}$$

The *wave length* is the value of z where $\lambda z = 2\pi$, or

$$\text{wave length} = 2\pi \sqrt{\frac{2D}{n}} \qquad \text{(E4.2.11)}$$

Finally, the *wave period* is the value of t corresponding to $nt = 2\pi$, or

$$\text{wave period} = \frac{2\pi}{n} \qquad \text{(E4.2.12)}$$

Now, we will attack the problem of different post hole depths. In St. Paul, Minnesota, the average daily temperature can be approximated as

$$T(^\circ C) = 6 + 18 \cos \left[\frac{2\pi}{365} (t - 113) \right] \qquad (E4.2.13)$$

where t is time in Julian days, and, in Boston, Massachusetts, the average daily surface temperature can be approximated by

$$T(^\circ C) = 9.5 + 12.5 \cos \left[\frac{2\pi}{365} (t - 113) \right] \qquad (E4.2.14)$$

In addition, the thermal diffusivity of the soil at both locations varies from $\alpha = 10^{-8}$ to $\alpha = 10^{-7}$ m^2/s. The post hole depth must be below the frost line, so that we do not have frost heave on our posts. The frost line would be at the depth where temperature never goes below freezing, $T = 0^\circ C$. We should set the cosine term from equation (E4.2.2) to -1 and set the temperature to $0^\circ C$ to determine this depth:

$$z_f = -\frac{1}{\lambda} \ln \left(\frac{T_a}{\Delta T} \right) = -\sqrt{\frac{2\alpha}{n}} \ln \left(\frac{T_a}{\Delta T} \right) \qquad (E4.2.15)$$

where $n = 2\pi/365$ days$^{-1} = 1.99 \times 10^{-7}$ s. Using the higher extreme in the range of thermal diffusion coefficients:

$$z_f(\text{St. Paul}) = -\sqrt{\frac{2(10^{-7} \text{ m}^2/\text{s})}{1.99 \times 10^{-7} \text{ s}}} \ln \left(\frac{6^\circ}{18^\circ} \right) = 1.09 \text{ m} \qquad (E4.2.16)$$

In Boston, equation (E4.4.15) gives

$$z_f(\text{Boston}) = -\sqrt{\frac{2(10^{-7} \text{ m}^2/\text{s})}{1.99 \times 10^{-7} \text{ s}}} \ln \left(\frac{9.5^\circ}{12.5^\circ} \right) = 0.27 \text{ m} \qquad (E4.2.17)$$

Comparison of equations (E4.2.16) and (E4.2.17) explains the origin of the difference in post hole depth between Boston and St. Paul. In Boston, post hole depth is that required for structural stability. In St. Paul, it is about avoiding frost heave. It is also interesting to note from Figure E4.2.1 that the temperature fluctuations at a depth of 1 m are damped by $1/e$, or 0.37, and the lag in the peak of temperature is $t = \lambda z/n$ or $1/(2\pi)$ years. The minimum temperature at a depth of 1 m appears almost 2 months after it is present at the surface.

B. Momentum Transport

The momentum of a fluid element per unit volume is given as

$$\text{Momentum/Volume} = \rho \vec{V} \qquad (4.5)$$

where $\vec{V}$ is a vector quantity (i.e., it has a direction) indicating velocity. The momentum experienced in a direction normal to the $\vec{V}$ vector is zero. The direction of the momentum vector, however, can change, complicating the analysis of momentum.

What is typically done is to divide momentum into component directions corresponding to the major axes. An example would be keeping track of momentum components in the x-direction and the y-direction separately.

Momentum can be transported in an analogous manner to mass or heat. The diffusion of momentum is described with a kinematic viscosity, v, which has SI units of m^2/s:

$$v = \frac{\mu}{\rho} \tag{4.6}$$

where μ is the viscosity of the fluid (g/m/s or N*s/m). Again, applying a similar operation as we did for mass on the rectangular control volume, equation (2.14) becomes the momentum transport equation:

$$\frac{\partial \rho \vec{V}}{\partial t} + \frac{\partial}{\partial x}\left(\rho u \vec{V}\right) + \frac{\partial}{\partial y}\left(\rho v \vec{V}\right) + \frac{\partial}{\partial z}\left(\rho w \vec{V}\right)$$

$$= \left[\frac{\partial}{\partial x}\left(v \frac{\partial \rho \vec{V}}{\partial x}\right) + \frac{\partial}{\partial y}\left(v \frac{\partial \rho \vec{V}}{\partial y}\right) + \frac{\partial}{\partial z}\left(v \frac{\partial \rho \vec{V}}{\partial z}\right)\right] + S \tag{4.7}$$

where S is the momentum source or sink rate per unit volume. Two common source/sink rates are pressure gradient and the force of gravity, where

$$S = \rho \vec{g} - \nabla \bullet P \tag{4.8}$$

where $\vec{g}$ is the acceleration of gravity in the direction of the vector $\vec{g}$, P is fluid pressure, and ∇ is a gradient operator in the direction of the vector $\vec{V}$. If the flow is incompressible, the resulting equation is analogous to equation (2.18):

$$\frac{\partial \vec{V}}{\partial t} + u\frac{\partial \vec{V}}{\partial x} + v\frac{\partial \vec{V}}{\partial y} + w\frac{\partial \vec{V}}{\partial z} = v\left(\frac{\partial^2 \vec{V}}{\partial x^2} + \frac{\partial^2 \vec{V}}{\partial y^2} + \frac{\partial^2 \vec{V}}{\partial z^2}\right) + \vec{g} - \frac{1}{\rho}\nabla \bullet P \tag{4.9}$$

Since the decision was made to work in Cartesian coordinates (our box control volume), in developing equations (2.14) and (2.18), we might as well divide the vectors into Cartesian coordinates:

$$\vec{V} = \vec{i}u + \vec{j}v + \vec{k}w \tag{4.10}$$

$$\vec{g} = \vec{i}g_x + \vec{j}g_y + \vec{k}g_z \tag{4.11}$$

$$\nabla = \vec{i}\frac{\partial}{\partial x} + \vec{j}\frac{\partial}{\partial y} + \vec{k}\frac{\partial}{\partial z} \tag{4.12}$$

where $\vec{i}$, $\vec{j}$, and $\vec{k}$ correspond to the x-, y-, and z-coordinates, respectively, and u, v, w and g_x, g_y, g_z are the x-, y-, and z-components of the velocity and gravity vectors, respectively. Then, equation (4.9) can be expressed as three equations without vector notation:

i-component of momentum

$$\frac{\partial u}{\partial t} + u\frac{\partial u}{\partial x} + v\frac{\partial u}{\partial y} + w\frac{\partial u}{\partial z} = v\left(\frac{\partial^2 u}{\partial x^2} + \frac{\partial^2 u}{\partial y^2} + \frac{\partial^2 u}{\partial z^2}\right) + g_x - \frac{1}{\rho}\frac{\partial P}{\partial x} \tag{4.13}$$

j-component of momentum

$$\frac{\partial v}{\partial t} + u\frac{\partial v}{\partial x} + v\frac{\partial v}{\partial y} + w\frac{\partial v}{\partial z} = \nu\left(\frac{\partial^2 v}{\partial x^2} + \frac{\partial^2 v}{\partial y^2} + \frac{\partial^2 v}{\partial z^2}\right) + g_y - \frac{1}{\rho}\frac{\partial P}{\partial y} \qquad (4.14)$$

k-component of momentum

$$\frac{\partial w}{\partial t} + u\frac{\partial w}{\partial x} + v\frac{\partial w}{\partial y} + w\frac{\partial w}{\partial z} = \nu\left(\frac{\partial^2 w}{\partial x^2} + \frac{\partial^2 w}{\partial y^2} + \frac{\partial^2 w}{\partial z^2}\right) + g_z - \frac{1}{\rho}\frac{\partial P}{\partial z} \qquad (4.15)$$

These are called the incompressible Navier–Stokes equations in Cartesian coordinates.

We can use equation (2.21), some intuition, and equations (4.13), (4.14), and (4.15) to infer the Navier–Stokes equation in cylindrical coordinates:

$$\frac{\partial v_z}{\partial t} + \frac{v_r}{r}\frac{\partial}{\partial r}(r v_z) + \frac{v_\theta}{r}\frac{\partial v_z}{\partial \theta} + v_z\frac{\partial v_z}{\partial z} = \nu\left[\frac{1}{r}\frac{\partial}{\partial r}\left(r\frac{\partial v_z}{\partial r}\right) + \frac{1}{r^2}\frac{\partial^2 v_z}{\partial \theta^2} + \frac{\partial^2 v_z}{\partial z^2}\right]$$
$$+ g_z - \frac{1}{\rho}\frac{\partial P}{\partial z} \qquad (4.16)$$

$$\frac{\partial v_r}{\partial t} + \frac{v_r}{r}\frac{\partial}{\partial r}(r v_r) + \frac{v_\theta}{r}\frac{\partial v_r}{\partial \theta} + v_z\frac{\partial C}{\partial z} = \nu\left[\frac{1}{r}\frac{\partial}{\partial r}\left(r\frac{\partial v_r}{\partial r}\right) + \frac{1}{r^2}\frac{\partial^2 v_r}{\partial \theta^2} + \frac{\partial^2 v_r}{\partial z^2}\right]$$
$$+ g_r - \frac{1}{\rho}\frac{\partial P}{\partial r} \qquad (4.17)$$

$$\frac{\partial v_\theta}{\partial t} + \frac{v_r}{r}\frac{\partial}{\partial r}(r v_\theta) + \frac{v_\theta}{r}\frac{\partial v_\theta}{\partial \theta} + v_z\frac{\partial v_\theta}{\partial z} = \nu\left[\frac{1}{r}\frac{\partial}{\partial r}\left(r\frac{\partial v_\theta}{\partial r}\right) + \frac{1}{r^2}\frac{\partial^2 v_\theta}{\partial \theta^2} + \frac{\partial^2 v_\theta}{\partial z^2}\right]$$
$$+ g_\theta - \frac{1}{\rho}\frac{\partial P}{\partial \theta} \qquad (4.18)$$

where the cylindrical coordinate system and notation are illustrated in Figure 2.6.

EXAMPLE 4.3: *Hagan Poiseuille flow – laminar, steady incompressible flow in a long pipe with a linear pressure gradient (first-order, nonlinear solution to an ordinary differential equation)*

An incompressible fluid flows in a laminar steady fashion through a long pipe with a linear pressure gradient. Describe the velocity profile and determine the relationship for the Darcy–Weisbach friction factor.

A sketch of a pipe section is given in Figure E4.3.1, with the coordinate system and the vector notation. Typically, a length of 50 diameters from the inlet is required before a fully developed (long pipe) flow is assumed to be reached. The following can be assumed for this problem statement:

1. Steady: $\partial/\partial t = 0$
2. Laminar: $v_r = 0$
3. Long pipe: $v_\theta = 0$, $v_z \neq v_z(z, \theta)$

Figure E4.3.1. Pipe section for Example 4.3.

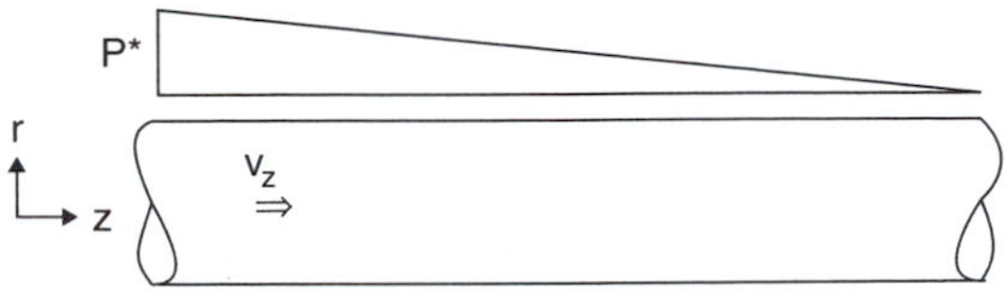

Now, equations (4.17) and (4.18) become meaningless because they result in $0 = 0$, and equation (4.16) becomes

$$\frac{\mu}{r}\frac{\partial}{\partial r}\left(r\frac{\partial v_z}{\partial r}\right) = \frac{1}{\rho}\frac{\partial P}{\partial z} - g_z \qquad \text{(E4.3.1)}$$

The right-hand side of equation (E4.3.1) is a constant because of the linear pressure gradient. The left-hand side can thus be written as an ordinary differential equation:

$$\frac{\mu}{r}\frac{d}{dr}\left(r\frac{dv_z}{dr}\right) = \frac{1}{\rho}\frac{\partial P}{\partial z} - g_z = \frac{\partial P^*}{\partial z} \qquad \text{(E4.3.2)}$$

where $P^* = P + \rho g\, El$ is the pressure minus the impact of fluid elevation on pressure through the hydrostatic pressure relationship. El is the elevation of the fluid element. Equation (E4.3.2) has the following boundary conditions:

1. At $r = 0$, $v_z = v_{z\max}$ from symmetry of the flow about the r-axis, which tells us that $dv_z/dr = 0$
2. At $r = R$, $u = 0$

We will consider the term in parentheses to be the variable of interest and integrate that variable with respect to r:

$$r\frac{dv_z}{dr} = \frac{\partial P^*/\partial z}{2\mu}r^2 + \beta_1 \qquad \text{(E4.3.3)}$$

Applying boundary condition 1 to equation (E4.3.3) gives $\beta_1 = 0$. Integrating v_z with respect to r gives

$$v_z = \frac{\partial P^*/\partial z}{4\mu}r^2 + \beta_2 \qquad \text{(E4.3.4)}$$

Applying boundary condition 2 to equation (E4.3.4):

$$0 = \frac{\partial P^*/\partial z}{4\mu}R^2 + \beta_2$$

or

$$\beta_2 = -\frac{\partial P^*/\partial z}{4\mu}R^2$$

Then, equation (E4.3.4) becomes

$$v_z = -\left(\frac{\partial P^*}{\partial z}\right)\frac{R^2}{4\mu}\left[1 - \left(\frac{r}{R}\right)^2\right] \qquad \text{(E4.3.5)}$$

Equation (E4.3.5) is a paraboloid, answering the first question of Example 4.3. To answer the second question, we need to investigate some characteristics of equation (E4.3.5).

The maximum velocity, v_{zmax}, occurs at $r = 0$, or

$$v_{zmax} = -\frac{\partial P^*}{\partial z}\frac{R^2}{4\mu} \tag{E4.3.6}$$

and the cross-sectional mean velocity is given by $\overline{v_z} = Q/A$, where Q is fluid discharge and A is cross-sectional area. Then,

$$Q = \int_0^r v_z\,(2\pi r)dr = -\frac{\pi R^4}{8\mu}\frac{dP^*}{dz} \tag{E4.3.7}$$

and

$$A = \pi R^2 \tag{E4.3.8}$$

gives

$$\overline{v_z} = -\frac{R^2}{8\mu}\frac{\partial P^*}{\partial z} = \frac{v_{zmax}}{2} \tag{E4.3.9}$$

Friction occurs due to shear at the walls (τ), where for the most common fluids the assumption proposed by Isaac Newton applies (Newtonism fluids):

$$\tau = \mu\frac{\partial v_z}{\partial r}\bigg|_{r=R} = \frac{R}{2}\frac{\partial P^*}{\partial z} = -\frac{4\mu\overline{v_z}}{R} \tag{E4.3.10}$$

where equations (E4.3.5) and (E4.3.9) have also been applied. The pressure gradient in a pipe may be computed by summing the shear stress that the wall feels, which is equal and opposite to that which the fluid feels:

$$\frac{\partial P^*}{\partial z} = \frac{2\tau}{R} \tag{E4.3.11}$$

Because the pressure gradient is linear, the change in pressure, ΔP^*, over a distance L can be found by integrating equation (E4.3.11):

$$\Delta P^* = \frac{2\tau L}{R} \tag{E4.3.12}$$

Then, substituting equation (E4.3.10) into equation (E4.3.12),

$$\Delta P^* = \frac{8\mu\overline{v_z}L}{R^2} \tag{E4.3.13}$$

Now, the Darcy–Weisbach equation for head loss defines the friction factor, f:

$$h_L = \frac{\Delta P^*}{\rho g} = f\frac{L}{d}\frac{\overline{v_z}^2}{2g} \tag{E4.3.14}$$

If we substitute equation (E4.3.12) or (E4.3.13) into (E4.3.14) and rearrange,

$$f = \frac{8\tau}{\rho\overline{v_z}^2} = \frac{32\mu}{\rho R\overline{v_z}} = \frac{64}{Re} \tag{E4.3.15}$$

where $Re = \overline{v_z}d/\nu$ is the Reynolds number and d is the pipe diameter.

If there is laminar flow in a pipe or tube, equation (E4.3.15) is used to compute the friction factor accurately after some velocity profile development length.

EXAMPLE 4.4: *Development of a momentum boundary layer over a solid surface (Blasius solution)*

Momentum boundary layer calculations are useful to estimate the skin friction on a number of objects, such as on a ship hull, airplane fuselage and wings, a water surface, and a terrestrial surface. Once we know the boundary layer thickness, occurring where the velocity is 99% of the free-stream velocity, skin friction coefficient and the skin friction drag on the solid surface can be calculated. Estimate the laminar boundary layer thickness of a 1-m-long, thin flat plate moving through a calm atmosphere at 20 m/s.

We have chosen a thin flat plate because the pressure gradient will be zero. With a curved body, there is a varying pressure gradient, and the solution to a given shape is more involved. We will orient the x-axis parallel to the plate and approaching velocity, and the z-axis normal to the plate. Our coordinate system will be moving with the plate, such that the plate sees air flowing past at 20 m/s. The following boundary conditions can then be applied:

1. At $x = 0, z \neq 0, u = U = 20$ m/s
2. At $x > 0, z = 0, u = 0$

We will be solving equation (4.13) for u and can make the following assumptions to the first order:

1. Steady state: $\partial/\partial t = 0$
2. Small y-velocity components: $u\, \partial u/\partial x \gg v\, \partial u/\partial y$
3. The change in velocity gradient with regard to x and y is small, compared with the change in gradient with respect to z: $\partial^2 u/\partial z^2 \gg \partial^2 u/\partial x^2, \partial^2 u/\partial y^2$
4. No gravitational force in the x-direction: $g_x = 0$
5. No pressure gradient: $\partial P/\partial x = 0$

Then, equation (4.15) becomes

$$u\frac{\partial u}{\partial x} + w\frac{\partial u}{\partial z} = v\frac{\partial^2 u}{\partial z^2} \qquad \text{(E4.4.1)}$$

with boundary conditions:

1. $x > 0, z = 0; u = 0$
2. $x = 0, z \neq 0; u = U$

Equation (E4.4.1) is a nonlinear partial differential equation, because of the velocity u that appears in front of the velocity gradient $\partial u/\partial x$. The boundary layer thickness is generally defined as the distance from the plate where the momentum reaches 99% of the free-stream momentum. We will assign (Blasius, 1908)

$$\eta = \frac{z}{x}\sqrt{Re_x} \qquad \text{(E4.4.2)}$$

where $Re_x = Ux/\nu$. We will introduce a stream function, ψ, that is constant along streamlines, with a gradient that is determined by the two equations

$$u = -\partial \psi / \partial y \tag{E4.4.3a}$$

and

$$v = \partial \psi / \partial x \tag{E4.4.3b}$$

If we assign

$$\psi = -\sqrt{\nu U x}\, f(\eta) \tag{E4.4.4}$$

then

$$u = U \frac{\partial f}{\partial \eta} \tag{E4.4.5a}$$

and

$$v = \frac{1}{2} \sqrt{\frac{\nu U}{x}} \left(\eta \frac{\partial f}{\partial \eta} - f \right) \tag{E4.4.5b}$$

Then, substituting equations (E4.4.5) into equation (E4.4.1) gives

$$2 \frac{d^3 f}{d\eta^3} + \eta \frac{d^2 f}{d\eta^2} = 0 \tag{E4.4.6}$$

Equation (E4.4.6) is an ordinary differential equation which may be solved numerically. First, we need to determine our boundary conditions in terms of η:

1. $x > 0, z = 0 \quad \eta = 0 \quad u = 0$
2. $x = 0, z \neq 0 \quad \eta = \infty \quad u = U$

The solution of equation (E4.4.6) with these boundary conditions is illustrated in Figure E4.4.1.

The transition to a turbulent boundary layer for a flat plate has been experimentally determined to occur at an Re_x value of between 3×10^5 and 6×10^5. For this example, the transition would occur between 15 and 30 cm after the start of the plate. Thus, the computations for a laminar boundary layer at 0.6 and 1 m are not realistic. However, the Blasius solution helps in the analysis of experimental data for a turbulent boundary layer, because it can tell us which parameters are likely to be important for this analysis, although the equations may take a different form.

The boundary layer thickness, δ, is the thickness, z, where $u/U = 0.99$. From the numerical solution, this occurs at $\eta = 4.85$. Then, the thickness of the boundary layer, δ, is found from a rearrangement of equation (E4.4.2) and is provided for different distances in equation (E4.4.7) and Table E4.4.1.

$$\delta = \frac{\eta x}{\sqrt{Re_x}} \tag{E4.4.7}$$

A local shear stress may be computed from the equation,

$$\tau_0 = \mu \left. \frac{\partial u}{\partial z} \right|_{z=0} = \frac{0.664}{\sqrt{Re_x}} \rho \frac{U^2}{2} \tag{E4.4.8}$$

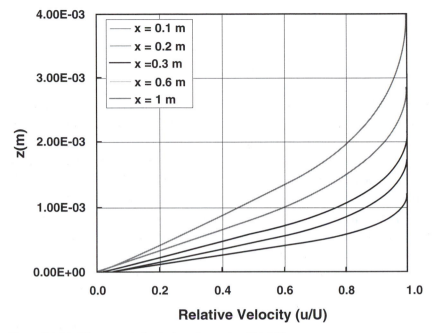

Figure E4.4.1. The numerical solution of equation (E4.4.6.).

which may be visualized as a flux of momentum into the plate. A skin friction coefficient is defined as $c_f = 2\tau_0/(\rho U^2)$. These values are also provided in Table E4.4.1. As the boundary layer thickness increases, the skin friction coefficient decreases, as one would expect.

C. Boundary Layer Analogies

It is apparent that the source and sink terms can be different between mass, heat, and momentum transport. There is another significant difference, however, related to the magnitude of the "diffusion coefficient" for mass, heat, and momentum.

$$\frac{\text{Momentum flux rate}}{\text{Unit volume}} = -\nu \frac{\partial(\rho V)}{\partial z} \tag{4.19}$$

in the heat transport equation,

$$\frac{\text{Heat flux rate}}{\text{Unit volume}} = \frac{-k_T}{\rho \, C_p} \frac{\partial T}{\partial z} = \alpha \frac{\partial T}{\partial z} \tag{4.20}$$

Table E4.4.1: *Boundary layer thickness and skin friction coefficient for a flat plate moving at 20 m/s in the atmosphere*

Distance (m)	0.1	0.2	0.3	0.6	1.0
Boundary layer thickness (m)	1.08E-03	1.53E-03	1.88E-03	3.25E-03	4.85E-03
Skin friction coefficient, c_f	1.48E-03	1.05E-03	8.57E-04	7.42E-04	6.64E-04

and in the mass transport equation,

$$\frac{\text{Mass flux rate}}{\text{Unit volume}} = -D\frac{\partial C}{\partial z} \tag{4.21}$$

The boundary conditions can be made to look similar through the conversion to dimensionless variables. These equations can be used to compute a momentum boundary layer thickness, such as in Example 4.4, a thermal boundary thickness, and a concentration boundary layer thickness. There are entire texts organized around these analogies, the best known being Bird, Stewart, and Lightfoot (1960). The primary difference within these analogies is shown in Table 4.1, where the momentum flux rate per momentum gradient, the heat flux rate per heat gradient, and the mass flux per mass gradient are given. This will provide a good comparison if the boundary conditions are made nondimensional to vary between 0 and 1.

It is seen that we are comparing kinematic viscosity, thermal diffusivity, and diffusivity of the medium for both air and water. In air, these numbers are all of the same order of magnitude, meaning that air provides a similar resistance to the transport of momentum, heat, and mass. In fact, there are two dimensionless numbers that will tell us these ratios: the Prandtl number ($Pr = \rho C_p \nu / k_T = \nu/\alpha$) and the Schmidt number ($Sc = \nu/D$). The Prandtl number for air at $20°C$ is 0.7. The Schmidt number for air is between 0.2 and 2 for helium and hexane, respectively. The magnitude of both of these numbers are on the order of 1, meaning that whether it is momentum transport, heat transport, or mass transport that we are concerned with, the results will be on the same order once the boundary conditions have been made dimensionless.

Considering the flux of momentum, heat, and mass in water, we can see that, although these are similar problems, the results could be quite different. The Prandtl number of 7 in water means that momentum flux per momentum gradient through the viscous sublayer operates with seven times the analogous diffusion of heat flux per thermal gradient. A Schmidt number of 160 to 1,600 in water indicates that momentum flux per momentum gradient through the viscous sublayer is 160 to 1,600 greater than mass flux per mass gradient. The result is that the growth of the viscous layer thickness will be much faster than that of the thermal and concentration boundary layer thicknesses.

Table 4.1: *Comparison of momentum, heat, and mass flux per gradient for air and water*

	Air ($20°C$)	Water ($20°C$)
Momentum flux/gradient $= \nu$	1.5×10^{-5} m^2/s	1.0×10^{-6} m^2/s
Heat flux/gradient $= k/(\rho C_p)$	2.0×10^{-5} m^2/s	1.5×10^{-7} m^2/s
Mass flux/gradient $= D$	$0.8–7.0 \times 10^{-5}$ m^2/s	$0.6–6.0 \times 10^{-9}$ m^2/s
Prandtl number, $Pr = \rho C_p \nu / k$	0.7	7
Schmidt number, $Sc = \nu/D$	0.2–2	160–1,600

While the viscous sublayer may be important for momentum transport, it is *everything* for mass and heat transport through liquids. Virtually the entire concentration boundary layer is within the viscous sublayer! This difference is important in our assumptions related to interfacial transport, the topic of Chapter 8, where mass is transported through an interfacial boundary layer.

D. Similitude and Transport Experiments

There are many transport conditions where experiments are needed to determine coefficients to be used in the solution. Examples are an air–water transfer coefficient, a sediment–water transfer coefficient, and an eddy diffusion coefficient. These coefficients are usually specific to the type of boundary conditions and are determined from empirical characterization relations. These relations, in turn, are based on experimental data.

So, how does an experiment run in a laboratory at different velocities and with different length scales relate to the problem that needs to be solved? A powerful tool to utilize in this task is similitude. By choosing an appropriate length scale and velocity scale, dimensionless parameters can be developed to characterize the experiments and to apply the results to other "similar" applications.

There are three techniques of developing the dimensionless similarity parameters. The use of Buckingham's pi theorem can be found in most fluid mechanics books, where the variables of importance are used to determine the number of dimensionless parameters that should describe an application and help to identify these parameters. One difficulty with Buckingham's pi theorem is the unspecified form of the dimensionless numbers, which can result in unusual combinations of parameters.

The second technique is physical insight into the problem, where ratios of forces or mass/heat transport determinants are factored to develop dimensionless numbers. This technique can also be found in most fluid mechanics texts.

The third technique uses the fundamental equations, and converts them into dimensionless equations with length and velocity scales important to the application. From these dimensionless equations, dimensionless numbers will evolve. It is this technique that will be described herein.

1. Dimensionless Mass Transport Equation

We start with equation (2.18):

$$\frac{\partial C}{\partial t} + u\frac{\partial C}{\partial x} + v\frac{\partial C}{\partial y} + w\frac{\partial C}{\partial z} = D\left(\frac{\partial^2 C}{\partial x^2} + \frac{\partial^2 C}{\partial y^2} + \frac{\partial^2 C}{\partial z^2}\right) + S \qquad (4.18)$$

where a typical relation for S is

$$S = k(C - C_E) \qquad (4.22)$$

and k is a first-order rate of reaction coefficient. C_E is an equilibrium concentration that will result from the transfer after a long time.

Let us define some dimensionless variables:

$$\widetilde{C} = C/C_E$$
$$X = x/L_x$$
$$Y = y/L_y$$
$$Z = z/L_z$$
$$U = u/U_r$$
$$V = v/V_r$$
$$W = w/W_r$$
$$\widetilde{T} = tU_r/L_x$$

where L_x, L_y, and L_z are important length dimensions and U_r, V_r, and W_r are important velocities in the x-, y-, and z-directions, respectively. We would like to choose these dimensionless variables so that the limits of each are 0 and 1. Now, we can substitute these dimensionless variables for the dimensional ones ($C \Rightarrow C_E\widetilde{C}$, $x \Rightarrow XL_x$, etc.) in equation (2.18). After dividing by D and C_E and multiplying by L_x^2, the following equation results:

$$Pe_x\frac{\partial \widetilde{C}}{\partial \widetilde{T}} + Pe_x U\frac{\partial \widetilde{C}}{\partial X} + Pe_y\frac{L_x^2}{L_y^2}V\frac{\partial \widetilde{C}}{\partial Y} + Pe_z\frac{L_x^2}{L_z^2}W\frac{\partial \widetilde{C}}{\partial Z}$$

$$= \left(\frac{\partial^2 \widetilde{C}}{\partial X^2} + \frac{L_x^2}{L_y^2}\frac{\partial^2 \widetilde{C}}{\partial Y^2} + \frac{L_x^2}{L_z^2}\frac{\partial^2 \widetilde{C}}{\partial Z^2}\right) + N_R\frac{L_x}{L}\left(\widetilde{C} - 1\right) \qquad (4.23)$$

where $Pe_x = U_r L_x/D$, $Pe_y = V_r L_y/D$, and $Pe_z = W_r L_z/D$ are Peclet numbers and $N_R = k L_x/D$ is a dimensionless reaction number, which could also be a Sherwood number if k is an interface mass transfer rate coefficient. Since the dimensionless variables in equation (4.23) are all based on reference variables that would occur at different scales, the important parameters to match between the laboratory experiments and the field application are the Peclet numbers, the length ratios, and the reaction number or Sherwood number. In addition, an order of magnitude analysis can eliminate many of the terms in equation (4.23), depending on the application.

EXAMPLE 4.5: *Laboratory experiments on mass transfer from sediments (similitude in mass transport)*

You are engaged in laboratory flume experiments on transfer of dissolved oxygen into the sediments below the flowing water. The goal is to measure the sediment-water mass transfer coefficient and relate it to other parameters of the flow field. The flume is 20 m in length, with a depth between 3 and 10 cm and velocity between

0.05 and 0.5 m/s. Which dimensionless parameters would be best to characterize your experiments, so that they can be applied in field applications?

We will start with equation (4.23) and perform an order of magnitude analysis to reduce the terms to those that are important to this situation. First, the transfer is from the flowing water into the sediments, so the appropriate "width of the flow" would be the depth of the flow, h. We will orient the z-direction so that it is vertical upward, and then the L_z characteristic length scale would also be h.

The characteristic velocity scales are typically the mean velocities in each direction, since they are known quantities. Turbulence can be characterized by the largest of these, since that will drive the velocity of the turbulent eddies. The mean velocities in the y-direction (transverse) and z- direction (vertical) are zero, so the third and fourth terms on the left-hand side of equation (4.23) become zero.

The flux is in the z-direction, so the gradients in the z-direction are much greater than those in the x- and y-directions. The order-of-magnitude analysis of the second derivatives typically follows those of the first derivatives, so the first two terms in brackets are much smaller than the third term.

Then, for this application, equation (4.23) becomes

$$Pe_x \frac{\partial \widetilde{C}}{\partial \widetilde{T}} + Pe_x U \frac{\partial \widetilde{C}}{\partial X} = \frac{L_x^2}{L_z^2} \frac{\partial^2 \widetilde{C}}{\partial Z^2} + Sh \frac{L_x}{L_z} (\widetilde{C} - 1) \qquad (E4.5.1)$$

The interaction of flow and diffusion in these terms is typically not straightforward, so the Peclet number is typically divided into $Re_x Sc$, where Re_x is a Reynolds number $(U_r L_x / \nu)$ and Sc is the Schmidt number (ν/D). The width of the flow to be used in the Sherwood number for this application is L_z. We will therefore multiply all terms by L_z/L_x:

$$Re Sc \frac{\partial \widetilde{C}}{\partial \widetilde{T}} + Re Sc U \frac{\partial \widetilde{C}}{\partial X} = \frac{h}{L_T} \frac{\partial^2 \widetilde{C}}{\partial Z^2} + Sh \frac{h}{L_T} (\widetilde{C} - 1) \qquad (E4.5.2)$$

where $Re = \overline{U}h/\nu$, $Sh = Kh/D$, $\overline{U}$ is the cross-sectional mean velocity in the flume, h is channel depth, K is the sediment–water mass transfer coefficient, and L_T is the length of the test section or the region of application.

Equation (E4.5.2) can be used to identify the important dimensionless parameters for extrapolation of the experimental results to field applications:

$$Sh = F(Re, Sc, h/L_T) \qquad (E4.5.3)$$

This type of relationship is often written as

$$Sh = \beta_0 Sc^{\beta_1} Re^{\beta_2} \left(\frac{h}{L_T}\right)^{\beta_3} \qquad (E4.5.4)$$

where $\beta_0, \beta_1, \beta_2,$ and β_3 are coefficients to be determined by log-linear regression (take the log of both sides of equation (E4.5.4) and perform a linear regression on the log terms). For this case, the importance of h/L_T will diminish as L_T becomes large.

Table 4.2: *Relations that characterize mass transfer coefficients for various applications (Cussler, 1997)*

Physical situation	Basic equation*†	Key variables	Remarks
Solid interfaces			
Membrane	$\dfrac{kl}{D} = 1$	l = membrane thickness	Often applied even where membrane is hypothetical
Laminar flow along flat plate	$\dfrac{kz}{D} = 0.323 \left(\dfrac{zv^0}{\nu}\right)^{1/2} \left(\dfrac{\nu}{D}\right)^{1/3}$	z = distance from start of plate v^0 = bulk velocity	Solid theoretical foundation
Turbulent flow through horizontal slit	$\dfrac{kd}{D} = 0.026 \left(\dfrac{dv^0}{\nu}\right)^{0.8} \left(\dfrac{\nu}{D}\right)^{1/3}$	v^0 = average velocity in slit $d = (2/\pi)$ (slit width)	Mass transfer here is identical with that in a pipe of equal wetted perimeter
Turbulent flow through circular pipe	$\dfrac{kd}{D} = 0.026 \left(\dfrac{dv^0}{\nu}\right)^{0.8} \left(\dfrac{\nu}{D}\right)^{1/3}$	v^0 = average velocity in pipe d = pipe diameter	Same as slit, because only wall region is involved
Laminar flow through circular pipe‡	$\dfrac{kd}{D} = 1.86 \left(\dfrac{dv^0}{D}\right)^{0.8}$	d = pipe diameter L = pipe length v^0 = average velocity in pipe	Not reliable when $(dv/D) < 10$ because of free convection
Forced convection around a solid sphere	$\dfrac{kd}{D} = 2.0 + 0.6 \left(\dfrac{dv^0}{\nu}\right)^{1/2} \left(\dfrac{\nu}{D}\right)^{1/3}$	d = sphere diameter v^0 = velocity of sphere	Very difficult to reach $(kd/D) = 2$ experimentally; no sudden laminar–turbulent transition
Free convection around a solid sphere	$\dfrac{kd}{D} = 2.0 + 0.6 \left(\dfrac{d^3[\Delta\rho]g}{\rho\nu^2}\right)^{1/2} \left(\dfrac{\nu}{D}\right)^{1/3}$	d = sphere diameter g = gravitational acceleration	For a 1-cm sphere in water, free convection is important when $\Delta\rho = 10^{-9}$ g/cm^3
Spinning disc	$\dfrac{kd}{D} = 0.62 \left(\dfrac{d^2 w}{\nu}\right)^{1/2} \left(\dfrac{\nu}{D}\right)^{1/3}$	d = disc diameter w = disc rotation (radians/time)	Valid for Reynolds numbers between 100 and 20,000
Flow normal to capillary bed	$\dfrac{kd}{D} = f\left(\dfrac{dv^0}{\nu}, \dfrac{\nu}{D}\right)$	d = tube diameter v^0 = average velocity	Large number of correlations with different exponents found by analogy with heat transfer
Packed beds	$\dfrac{kd}{D} = 1.17 \left(\dfrac{dv^0}{\nu}\right)^{0.58} \left(\dfrac{\nu}{D}\right)^{1/3}$	d = particle diameter v^0 = superficial velocity	The superficial velocity is that which would exist without packing

Physical situation	Basic equation*†	Key variables	Remarks
Fluid–fluid interfaces			
Drops or bubbles in stirred solution	$\dfrac{kL}{D} = 0.13 \left(\dfrac{L^2(P/v)}{pv^3} \right)^{3/4} \left(\dfrac{v}{D} \right)^{1/3}$	$L = $ stimer length $P/V = $ power per volume	Correlation versus power per volume are common for dispersions
Large drops in unstirred solution	$\dfrac{kd}{D} = 0.42 \left(\dfrac{d^3 \Delta \rho g}{pv^2} \right)^{1/3} \left(\dfrac{v}{D} \right)^{1/2}$	$d = $ bubble diameter $\Delta \rho = $ density difference between bubble and surrounding fluid	"Large" is defined as $\sim$0.3-cm diameter
Small drops of pure solute in unstirred solution	$\dfrac{kd}{D} = 1.13 \left(\dfrac{dv^0}{D} \right)^{0.8}$	$d = $ bubble diameter $v^0 = $ bubble velocity	These behave like rigid spheres
Falling film	$\dfrac{kz}{D} = 0.69 \left(\dfrac{zv^0}{D} \right)^{1/2}$	$z = $ position along films $v^0 = $ average film velocity	Frequency embroidered and embellished

* Symbols used include the following: ρ is the fluid density; v is the kinematic viscosity; D is the diffusion coefficient of the material being transferred; and k is the local mass transfer coefficient. Other symbols are defined for the specific situation.

† The dimensionless groups are defined as follows: (dv/v) and (d^2w/v) are the Reynolds number; v/D is the Schmidt number; $(d^2 \Delta \rho g/\rho v^2)$ is the Grashöf number; kd/D is the Sherwood number; and k/v is the Stanton number.

‡ The mass transfer coefficient given here is the value averaged over the length.

Source: Data from Calderbank (1967), McCabe and Smith (1975), Schlichting (1979), Sherwood et al. (1975), and Treybal (1980).

A list of mass transfer coefficient relations provided in Table 4.2 illustrates the types of experimentally determined relations that exist. They are typically of the form of equation (E4.5.4), although some deviations occur due to a theoretical analysis of mass transfer. Theoretical and experimental analyses have shown that $Sh \sim Sc^{1/2}$ for an interface that acts like a fluid and $Sh \sim Sc^{1/3}$ for an interface that acts like a solid.

2. Dimensionless Heat Transport Equation

For heat transport in incompressible flow, we will start with equation (4.3):

$$\frac{\partial T}{\partial t} + u \frac{\partial T}{\partial x} + v \frac{\partial T}{\partial y} + w \frac{\partial T}{\partial z} = \alpha \left(\frac{\partial^2 T}{\partial x^2} + \frac{\partial^2 T}{\partial y^2} + \frac{\partial^2 T}{\partial z^2} \right) + \frac{S}{\rho C_p} \qquad (4.3)$$

and use similar dimensionless variables, except we now use a dimensionless temperature,

$$\theta = T/\Delta T$$

where ΔT is an appropriate temperature difference. Then, equation (4.3) becomes

$$Pe_x \frac{\partial \theta}{\partial \widetilde{T}} + Pe_x U \frac{\partial \theta}{\partial X} + Pe_y \frac{L_x^2}{L_y^2} V \frac{\partial \theta}{\partial Y} + Pe_z \frac{L_x^2}{L_z^2} W \frac{\partial \theta}{\partial Z}$$

$$= \left(\frac{\partial^2 \theta}{\partial X^2} + \frac{L_x^2}{L_y^2} \frac{\partial^2 \theta}{\partial Y^2} + \frac{L_x^2}{L_z^2} \frac{\partial^2 \theta}{\partial Z^2} \right) + \frac{L_x^2}{\alpha} \frac{S}{\rho C_p \Delta T} \quad (4.24)$$

where the Peclet numbers now use thermal diffusion coefficient, α, instead of the mass diffusion coefficient.

EXAMPLE 4.6: *Measurements of near-field thermal plume downstream of a power plant (similitude in heat transport)*

Field measurements of a thermal plume downstream of a thermal power plant, illustrated in Figure E4.6.1, are required to determine the impact of the heated water on the river biota. These are near field because they occur before the river is mixed across its width and depth. Since field measurements are expensive and time consuming, it is desirable to select the measurements on which to concentrate the effort.

Begin with equation (4.21). The power plant will be running at steady state, and the measurements will be taken at a low flow, such that $\partial \theta / \partial \widetilde{T}$ should be close to zero. We will then place the x-coordinate along the longitudinal directions of the river, so that the mean velocities in the y- and z-direction are zero. Finally, the source term in a river system is often approximated by (Brady et al., 1969)

$$S = \rho C_p K (T - T_E) \quad (E4.6.1)$$

where K is a heat transfer coefficient, incorporating the effects of wind and the non-linearity of long-wave radiation, and T_E is an equilibrium temperature, where net heat transfer with the atmosphere is zero, incorporating the impact of air temperature, humidity, long-wave radiation, and solar radiation. Then, equation (4.24) becomes

$$U \frac{\partial \theta}{\partial X} = \frac{1}{Pe_x} \left(\frac{\partial^2 \theta}{\partial X^2} + \frac{L_x^2}{L_y^2} \frac{\partial^2 \theta}{\partial Y^2} + \frac{L_x^2}{L_z^2} \frac{\partial^2 \theta}{\partial Z^2} \right) + \frac{L_x}{U_r} K (\theta - \theta_E) \quad (E4.6.2)$$

where $\theta_E = T_E / \Delta T$. Because the reference velocity will be the mean velocity, the diffusion coefficient in the Peclet number will be a sum of the turbulent diffusion

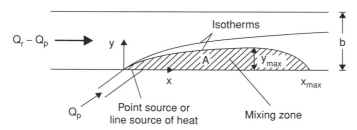

Figure E4.6.1. Illustration of isotherms (constant temperature lines) below a thermal power plant discharging to a river.

coefficient and the thermal diffusion coefficient. This procedure will be developed further in Chapter 5. We will assign the following parameters:

$U_r = Q/A$, the cross-sectional mean velocity

$\Delta T = $ the difference between the temperature of the inflow from the power plant and the ambient temperature

$L_y = $ mean river width, b

$L_z = $ mean river depth

Typically, the temperature is mixed over the depth quickly, such that $\partial^2\theta/\partial Z^2$ does not play a big role in river isotherms. The only difficult independent parameter to determine is L_x. We could assign this to be the length of the mixing zone, but it changes over discharge and other boundary conditions. We will, instead, assign this to be the width of the river, $L_x = L_y$, because the width has much to do with the length of the mixing zone. Then, equation (E4.6.2) becomes

$$U\frac{\partial\theta}{\partial X} = \frac{1}{Pe}\left(\frac{\partial^2\theta}{\partial X^2} + \frac{\partial^2\theta}{\partial Y^2}\right) + St(\theta - \theta_E) \tag{E4.6.3}$$

where $Pe = Qb/(A\alpha)$ and $St = K\,b\,A/Q$ is a Stanton number. Equation (E4.6.3) indicates that we need to be as accurate as possible with measurements of discharge, cross-sectional area, temperature difference at the plant outlet, mean river width, and the temperature and x,y-location of each measurement.

3. Dimensionless Momentum Transport Equation

We will consider one component of the Navier–Stokes equation,

$$\frac{\partial u}{\partial t} + u\frac{\partial u}{\partial x} + v\frac{\partial u}{\partial y} + w\frac{\partial u}{\partial z} = \nu\left(\frac{\partial^2 u}{\partial x^2} + \frac{\partial^2 u}{\partial y^2} + \frac{\partial^2 u}{\partial z^2}\right) + g_x - \frac{1}{\rho}\frac{\partial P}{\partial x} \tag{4.13}$$

and make it dimensionless by adding two dimensionless variables to those used previously:

$$\widetilde{P} = P/P_r$$
$$H = (h + El_0)/L$$

where P_r is a reference pressure, El_0 is the elevation of the channel bottom, and L is some reference length. We will also convert the x-component of gravitational force per unit mass, g_x, to that used for a water surface:

$$g_x = -g\frac{d(h + El_0)}{dx} \tag{4.25}$$

Then, equation (4.13) becomes:

$$\frac{\partial U}{\partial \widetilde{T}} + U\frac{\partial U}{\partial X} + \frac{V_r}{U_r}\frac{L_x}{L_y}V\frac{\partial U}{\partial Y} + \frac{W_r}{U_r}\frac{L_x}{L_z}W\frac{\partial U}{\partial Z}$$

$$= \frac{1}{Re}\left(\frac{\partial^2 U}{\partial X^2} + \frac{L_x^2}{L_y^2}\frac{\partial^2 U}{\partial Y^2} + \frac{L_x^2}{L_z^2}\frac{\partial^2 U}{\partial Z^2}\right) - \frac{1}{Fr^2}\frac{dH}{dX} - \frac{2}{Eu}\frac{\partial \widetilde{P}}{\partial X} \tag{4.26}$$

where $Re = U_r L_x/v$ is a Reynolds number, $Fr = U_r/\sqrt{gL}$ is a Froude number, and $Eu = P_r/(\rho U_r^2)$ is an Euler number.

From equation (4.26), it is seen that the important dimensionless parameters driving momentum transport are the Reynolds number, the Froude number, the Euler number, and the length and velocity ratios in the flow field. The dimensionless variables all vary between zero and a value close to one, so they are not significant in determining which terms in the governing equation are important.

EXAMPLE 4.7: *Measurements of the velocity profile in an open-channel flow (similitude in momentum transport)*

You will be measuring velocity profiles in a fully developed open-channel flow (no change with longitudinal distance). To get an idea of which parameters you need to measure accurately, you need to perform an order-of-magnitude analysis on equation (4.26).

The important length and velocity scales are as follows:

$L_y = h$, Q/A, the channel depth
$L_z = b$, the channel width
$L_x =$ the length of the flume
$U_r = Q/A$ the mean velocity in the longitudinal direction
$V_r = 0$
$W_r = 0$

Since the mean velocities in the y- and z-directions are zero, we will set the reference velocities in these directions to zero.

In addition, the appropriate length scale for the Froude number term would seem to be the depth of the channel. Then, equation (4.26) becomes

$$\frac{\partial U}{\partial \widetilde{T}} + U\frac{\partial U}{\partial X} = \frac{1}{Re}\left(\frac{\partial^2 U}{\partial X^2} + \frac{L^2}{h^2}\frac{\partial^2 U}{\partial Y^2} + \frac{L^2}{b^2}\frac{\partial^2 U}{\partial Z^2}\right) - \frac{1}{Fr^2}\frac{dH}{dX} - \frac{2}{Eu}\frac{\partial \widetilde{P}}{\partial X} \quad \text{(E4.7.1)}$$

The velocity profile in fully developed flow will be determined with the $\partial^2 U/\partial Y^2$ term in equation (E4.7.1). To perform these experiments, then, we must make sure that the following is true:

$$\frac{1}{Re}\frac{L^2}{h^2}\frac{\partial^2 U}{\partial Y^2} \gg \frac{\partial U}{\partial \widetilde{T}}, U\frac{\partial U}{\partial X} \quad \text{(E4.7.2)}$$

which will occur if L/h is sufficiently large. In addition, we need to make sure that

$$\frac{L^2}{h^2}\frac{\partial^2 U}{\partial Y^2} \gg \frac{\partial^2 U}{\partial X^2}, \frac{L^2}{b^2}\frac{\partial^2 U}{\partial Z^2} \quad \text{(E4.7.3)}$$

which will occur if L/h and b/h are sufficiently large. This can be used to explain why most research flumes have larger lengths and widths than depths. If the four conditions in equations (E4.7.2) and (E4.7.3) are proven in the experiments to be true, then equation (E4.7.1) becomes

$$\frac{1}{Re}\frac{L^2}{h^2}\frac{\partial^2 U}{\partial Y^2} = \frac{1}{Fr^2}\frac{dH}{dX} + \frac{2}{Eu}\frac{\partial \tilde{P}}{\partial X} \qquad \text{(E4.7.4)}$$

that can be rearranged to give

$$\frac{1}{Re_h}\frac{\partial^2 U}{\partial Y^2} = \frac{1}{Fr_h^2}\frac{dH}{dX} + \frac{h}{L}\frac{2}{Eu}\frac{\partial \tilde{P}}{\partial X} \qquad \text{(E4.7.5)}$$

where $Re_h = Q/(bv)$ and $Fr_h = Q/(gb^2h^3)^{1/2}$. If L/h is large, the third term in equation (E4.7.5) is generally small, because the boundary condition of atmospheric pressure always occurs at the free surface. Equation (E4.7.5) tells us that, if we want our measurements to be accurate, we should make sure that the conditions of equations (E4.7.2) and (E4.7.3) are met, and we need to accurately measure Q, h, b, and $d(h + El_0)/dx$, the water surface slope. Although this may seem obvious, equation (E4.7.5) puts the obvious answer in context of the fundamental transport equation. The result can also be used to place quantitative assessments on the required accuracy and the conditions of the experiment. This observation can save much time in the experiments.

E. Problems

1. Develop the heat transport equation in cylindrical coordinates for temperature.

2. Consider the steady, laminar flow of an incompressible fluid in a long and wide closed conduit channel subject to a linear pressure gradient. (a) Derive the equation for velocity profile. (b) Derive the equation for discharge per unit width and cross-sectional mean velocity, and compare this with the maximum velocity in the channel. (c) Derive the equation for wall shear stress on both walls and compare them. Explain the sign convention for shear stress on each wall.

3. A viscometer is an apparatus that measures the viscosity of a fluid. A common style consists of an outer fixed cylinder with an inner rotating cylinder. One that is being used has an outer cylinder radius of 15 cm, an inner radius of 14.25 cm, and a height of 22 cm. It takes a torque (force time radius) of 0.07 N-m to maintain an angular speed of 50 rpm at 4°C. What is the viscosity of the fluid that fills the annular region of the cylinders? What fluids could this be?

4. You are taking some experimental data on pipe friction and want the analyzed results to apply to other pipes and sizes. Using the dimensionless Navier–Stokes equations, determine the important dimensionless parameters in these experiments.

5. Develop the dimensionless Navier–Stokes equations for cylindrical coordinates.

6. Flow enters a pipe with some swirl, which is reduced over distance by pipe fiction. In an analysis of experiments on this process, apply the dimensionless

Navier–Stokes equations to the swirl reduction experiments and isolate dimensionless numbers that can be used to apply the experiments to various pipes and
discharges.

7. A large oscillating plate sits below an incompressible fluid with no horizontal
   pressure gradient. The speed of the oscillating plate is $U = U_0 \cos(nt)$, where
   U_0 and n are constant, and t is time. Assume laminar flow and that the time since
   the initiation of plate movement is large. Knowing the viscosity and density of
   the fluid, determine the velocity profile above the plate. What is the frequency
   and wave length of the velocity wave through the fluid?

8. Fluid flowing in the duct of problem 2, with a spacing of $2h$, where $h = 0.1$ m,
   has a wall temperature of 25°C. Assume that the flow is laminar, and find the
   centerline temperature at fluid velocities of 1, 5, and 20 m/s under the following
   conditions:

   a. Air flow with $k = 0.005$ W/(m °K)

   b. Water flow with $k = 0.6$ W/(m °K)

   c. Glycerin flow with $k = 0.3$ W/(m °K)

   Note that 1 W $= 1$ kg m^2/s^3.

5 Turbulent Diffusion

Turbulent diffusion is not really diffusion but the mixing of chemicals through turbulent eddies created by convection. Turbulent diffusion is thus a form of convection. Although it has the appearance of diffusion in the end (i.e., random mixing similar to diffusion), the causes of diffusion and turbulent diffusion are very different. Since the end products are similar, diffusion coefficients and turbulent diffusion coefficients are often simply added together. This process will be discussed in this chapter.

A. Background on Turbulent Flow

It is fairly safe to state that, except for flow through porous media, the environment experiences turbulent flow. The reason that we have not used a river, lake, or the atmosphere as an application in an example in Chapter 2 is that these flows are always turbulent. The example simply would not have been realistic. To emphasize this point, we will consider the constriction of a water or air flow that would be required to have the other option, laminar flow.

An experimentally based rule-of-thumb is that laminar flow often occurs when the pipe Reynolds number, Vd/ν, is less than 2,000, or when an open channel Reynolds number, Vh/ν, is less than 500, where V is the cross-sectional mean velocity, d is the pipe diameter, ν is the kinematic viscosity of the fluid, and h is the channel depth. The diameter or depth that would not be exceeded to have laminar flow by these experimental criteria is given in Table 5.1.

Table 5.1 shows that, with the boundary conditions present in most environmental flows (i.e., the Earth's surface, ocean top and bottom, river or lake bottom), turbulent flow would be the predominant condition. One exception that is important for interfacial mass transfer would be very close to an interface, such as air–solid, solid–liquid, or air–water interfaces, where the distance from the interface is too small for turbulence to occur. Because turbulence is an important source of mass transfer, the lack of turbulence very near the interface is also significant for mass transfer, where diffusion once again becomes the predominant transport mechanism. This will be discussed further in Chapter 8.

Table 5.1: *Maximum diameter or depth to have laminar flow, with the transition Reynolds number for a pipe at 2,000*

| V (m/s) | Water ($\nu = 10^{-6}\,\mathrm{m^2/s}$) | | Air ($\nu \sim 2 \times 10^{-5}\,\mathrm{m^2/s}$) |
	d (m)	h (m)	d (m)
10	2×10^{-4}	5×10^{-5}	0.004
3	7×10^{-4}	1.5×10^{-4}	0.014
1	0.002	0.0005	0.04
0.3	0.007	0.0015	0.14
0.1	0.02	0.005	0.4
0.03	0.07	0.015	1.4
0.01	0.2	0.05	4.0

What is turbulent flow? We will use the simple illustration of a free-surface flow given in Figure 5.1 to describe the essential points of the turbulence phenomena. Turbulent open-channel flow can be described with a temporal mean velocity profile that reaches a steady value with turbulent eddies superimposed on it. These turbulent eddies are continually moving about in three dimensions, restricted only by the boundaries of the flow, such that they are eliminated from the temporal mean velocity profile, $\overline{u}$ in Figure 5.1. It is this temporal mean velocity profile that is normally sketched in turbulent flows.

There will also be a temporal mean concentration. If there is a source or sink in the flow, or transport across the boundaries as in Figure 5.1, then the temporal mean concentration profile will eventually reach a value such as that given in Figure 5.1. This flux of compound seems to be from the bottom toward the top of the flow. Superimposed on this temporal mean concentration profile will be short-term variations in concentration caused by turbulent transport. The concentration profile is "flatter" in the middle of the flow because the large turbulent eddies that transport mass quickly are not as constrained by the flow boundaries in this region. Now, if we put a concentration–velocity probe into the flow at one location, the two traces of velocity and concentration versus time would look something like that shown in Figure 5.2.

It is convenient to divide the velocity and concentration traces into temporal mean values and fluctuating components:

$$u = \overline{u} + u' \tag{5.1}$$

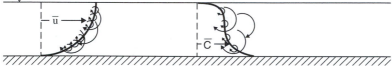

Figure 5.1. Turbulent eddies superimposed on a temporal mean velocity and temporal mean concentration profiles.

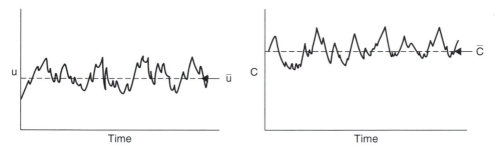

Figure 5.2. Time traces of typical measurements of velocity and concentration in a turbulent flow.

and

$$C = \overline{C} + C' \tag{5.2}$$

where $\overline{u}$ is the temporal mean velocity at a point location, u' is the fluctuating component of velocity (variable over time), $\overline{C}$ is the temporal mean concentration at a point location, and C' is the fluctuating concentration component of concentration that is also variable over time. A formal definition of $\overline{u}$ and $\overline{C}$ are as follows:

$$\overline{u} = \frac{1}{\Delta t} \int_0^{\Delta t} u \, dt \tag{5.3}$$

and

$$\overline{C} = \frac{1}{\Delta t} \int_0^{\Delta t} C \, dt \tag{5.4}$$

where Δt is long compared with the time period of the oscillating components.

B. Mass Transport Equation with Turbulent Diffusion Coefficients

In this section, we will derive the most common equations for dealing with mass transport in a turbulent flow. Beginning with equation (2.14), we will take the temporal mean of the entire equation and eventually end up with an equation that incorporates turbulent diffusion coefficients.

One of the conclusions of our consideration of the flux through a control volume was equation (2.14):

$$\frac{\partial C}{\partial t} + \frac{\partial (uC)}{\partial x} + \frac{\partial (vC)}{\partial y} + \frac{\partial (wC)}{\partial z} = \frac{\partial}{\partial x}\left(D\frac{\partial C}{\partial x}\right) + \frac{\partial}{\partial y}\left(D\frac{\partial C}{\partial y}\right) + \frac{\partial}{\partial z}\left(D\frac{\partial C}{\partial z}\right) + S \tag{2.14}$$

In a turbulent flow field, equation (2.14) is difficult to apply because C, u, v, and w are all highly variable functions of time and space. Osborne Reynolds (1895) reduced the complexities of applying equation (2.14) to a turbulent flow by taking the temporal mean of each term (e.g., the entire equation). Then, the mean value of a fluctuating component will be equal to zero, or

$$\overline{\frac{\partial C}{\partial t}} = \overline{\frac{\partial(\overline{C} + C')}{\partial t}} = \overline{\frac{\partial \overline{C}}{\partial t}} + \overline{\frac{\partial C'}{\partial t}} = \frac{\partial \overline{C}}{\partial t} + 0 \tag{5.5}$$

Equation (5.5), the change of a temporal mean over time, may seem like a misnomer, but it will be left in to identify changes in $\overline{C}$ over a longer time period than Δt. Continuing,

$$\overline{\frac{\partial C}{\partial x}} = \overline{\frac{\partial(\overline{C} + C')}{\partial x}} = \overline{\frac{\partial \overline{C}}{\partial x}} + \overline{\frac{\partial C'}{\partial x}} = \frac{\partial \overline{C}}{\partial x} \tag{5.6}$$

$$\overline{\frac{\partial C}{\partial y}} = \overline{\frac{\partial(\overline{C} + C')}{\partial y}} = \overline{\frac{\partial \overline{C}}{\partial y}} + \overline{\frac{\partial C'}{\partial y}} = \frac{\partial \overline{C}}{\partial y} \tag{5.7}$$

$$\overline{\frac{\partial C}{\partial z}} = \overline{\frac{\partial(\overline{C} + C')}{\partial z}} = \overline{\frac{\partial \overline{C}}{\partial z}} + \overline{\frac{\partial C'}{\partial z}} = \frac{\partial \overline{C}}{\partial z} \tag{5.8}$$

However, the temporal mean value of two fluctuating components, multiplied by each other, will not necessarily be zero:

$$\overline{u' C'} \neq \overline{u'}\,\overline{C'} \tag{5.9}$$

This is similar to a least-squares regression, where the mean error is zero, but the sum of square error is not. We will first deal with the x-component of our convective transport terms:

$$\overline{uC} = \overline{(\overline{u} + u')(\overline{C} + c')} = \overline{\overline{u}\overline{C}} + \overline{\overline{u}C'} + \overline{u'\overline{C}} + \overline{u'C'} \tag{5.10}$$

Three of the four terms in equation (5.10) may be reduced to something we know:

$$\overline{\overline{u}\overline{C}} = \overline{u}\,\overline{C} \tag{5.11}$$

$$\overline{\overline{u}C'} = 0 \tag{5.12}$$

$$\overline{u'\overline{C}} = 0 \tag{5.13}$$

but the fourth term will take some additional consideration, because it is not equal to zero:

$$\overline{u'\,C'} \neq 0 \tag{5.14}$$

By inference, we can now write the following for all three convective transport terms:

$$\overline{uC} = \overline{u}\overline{C} + \overline{u'C'} \tag{5.15}$$

$$\overline{vC} = \overline{v}\overline{C} + \overline{v'C'} \tag{5.16}$$

and

$$\overline{wC} = \overline{w}\overline{C} + \overline{w'C'} \tag{5.17}$$

Finally, applying continuity ($\overline{u} + \overline{v} + \overline{w} = 0$) to equation (2.14) and taking the temporal mean results of equations (5.5) to (5.8) and equations (5.15) to (5.17)

$$\frac{\partial \overline{C}}{\partial t} + \overline{u}\frac{\partial \overline{C}}{\partial x} + \overline{v}\frac{\partial \overline{C}}{\partial y} + \overline{w}\frac{\partial \overline{C}}{\partial z} = -\frac{\partial}{\partial x}\overline{u'C'} - \frac{\partial}{\partial y}\overline{v'C'} - \frac{\partial}{\partial z}\overline{w'C'}$$

$$+ \frac{\partial}{\partial x}\left(D\frac{\partial \overline{C}}{\partial x}\right) + \frac{\partial}{\partial y}\left(D\frac{\partial \overline{C}}{\partial y}\right) + \frac{\partial}{\partial z}\left(D\frac{\partial \overline{C}}{\partial z}\right) + \overline{S}$$

(5.18)

where we have moved the turbulent convective transport term to the right-hand side, because the concentration distribution that results from these terms looks similar to diffusion.

With this temporal mean process, we have reduced the terms for which we will have difficulty defining boundary conditions in turbulent flow fields from seven in equation (2.14) to three in equation (5.18). We will now deal with these three terms.

The diffusion equation is a useful and convenient equation to describe mixing in environmental flows, where the boundaries are often not easily defined. It also lends itself to analytical solutions and is fairly straightforward in numerical solutions. Although there is an alternative technique for solutions to mixing problems (the mixed cell method described in Chapter 6), there are complications of this alternative technique when applied to multiple dimensions and to flows that vary with space and time. Finally, we are comfortable with the diffusion equation, so we would prefer to use that to describe turbulent mixing if possible.

Therefore, let us consider the following thought process: if the end result of turbulence, when visualized from sufficient distance, looks like diffusion with seemingly random fluctuations, then we should be able to identify the terms causing these fluctuations in equation (5.18). Once we have identified them, we will relate them to a "turbulent diffusion coefficient" that describes the diffusion caused by turbulent eddies. Looking over the terms in equation (5.18) from left to right, we see an unsteady term, three mean convective terms, the three "unknown" terms, the diffusive terms, and the source/sink rate terms. It is not hard to figure out which terms should be used to describe our turbulent diffusion. The "unknown" terms are the only possibility.

In the late nineteenth century, Boussinesq (1877) probably went through something similar to the thought process described previously. The end result was the *Boussinesq eddy diffusion coefficient*:

$$-\overline{u'C'} = \varepsilon_x \frac{\partial \overline{C}}{\partial x}$$

(5.19a)

$$-\overline{v'C'} = \varepsilon_y \frac{\partial \overline{C}}{\partial y}$$

(5.19b)

$$-\overline{w'C'} = \varepsilon_z \frac{\partial \overline{C}}{\partial z}$$

(5.19c)

where ε_x, ε_y, and ε_z are the turbulent (or eddy) diffusion coefficients, with units of m^2/s similar to the (molecular) diffusion coefficients.

Then, equation (5.18) with equations (5.19a) to (5.19c) becomes

$$\frac{\partial \overline{C}}{\partial t} + \overline{u}\frac{\partial \overline{C}}{\partial x} + \overline{v}\frac{\partial \overline{C}}{\partial y} + \overline{w}\frac{\partial \overline{C}}{\partial z} = \frac{\partial}{\partial x}\left[(D + \varepsilon_x)\frac{\partial \overline{C}}{\partial x}\right]$$

$$+ \frac{\partial}{\partial y}\left[(D + \varepsilon_y)\frac{\partial \overline{C}}{\partial y}\right] + \frac{\partial}{\partial z}\left[(D + \varepsilon_z)\frac{\partial \overline{C}}{\partial z}\right] + \overline{S}$$

$$(5.20)$$

Turbulent diffusion is created by the flow field, which can vary with distance. Hence, eddy diffusivity cannot be assumed constant with distance. Removing that assumption leaves eddy diffusivity inside the brackets.

Character of Turbulent Diffusion Coefficients

A turbulent eddy can be visualized as a large number of different-sized rotating spheres or ellipsoids. Each sphere has subspheres and so on until the smallest eddy size is reached. The smallest eddies are dissipated by viscosity, which explains why turbulence does not occur in narrow passages: there is simply no room for eddies that will not be dissipated by viscosity.

The cause of the rotation is shear forces created by solid boundaries or variations in velocity lateral to the primary flow direction. A buoyant plume of smoke or steam, for example, will have a temporal mean velocity profile develop laterally to the plume, as the rising plume mixes with the ambient air. Turbulent eddies are formed by this velocity gradient and can be seen at the edge of the smoke or steam plume. The magnitude of turbulent diffusion coefficients is primarily dependent on the scale of turbulent eddies and the speed of the eddy rotation. As illustrated in Figure 5.3, a large eddy will have greater eddy diffusion coefficient than a small eddy, because it will transport a compound (or solute) farther in one rotation. Likewise, a faster spinning eddy will have a larger eddy diffusion coefficient than one that is the same

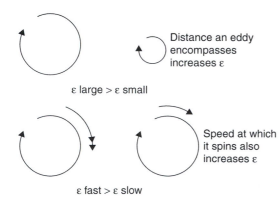

Figure 5.3. Character of turbulent diffusion coefficients.

size, but spinning more slowly because the solute simply gets there faster. These two facts provide meaning to the following observations:

1. The largest scale of turbulence is roughly equal to the smallest overall scale of the flow field. This may be seen in comparing the size of eddies at the edge of the smoke or steam plume to the width of the plume.
2. The rotational eddy velocity is roughly proportional to the velocity gradient times the eddy scale.
3. Eddy size decreases near boundaries to the flow field. Because the eddy size is zero at a solid boundary, and often close to zero at a fluid density interface (like an air–water interface), the turbulent eddy size has to decrease as one approaches the boundaries. In addition, because the flow cannot go through a boundary, the largest eddy size cannot be greater than the distance from the center of the eddy to the boundary.
4. Turbulent diffusion occurs because turbulent eddies are transporting mass, momentum, and energy over the eddy scale at the rotational velocity. This transport rate is generally orders of magnitude greater than the transport rate due to molecular motion. Thus, when a flow is turbulent, diffusion is normally ignored because $\varepsilon \gg D$. The exception is very near the flow boundaries, where the eddy size (and turbulent diffusion coefficient) decreases to zero.

Thus, what influences the velocity and scale of eddies? For the most part, it is the velocity gradients and scale of the flow. Velocity gradients are the change in velocity over distance. If we have a high velocity, we typically have a large velocity gradient somewhere in the flow field. At solid walls, for example, the velocity must go to zero. Thus, *the large velocity difference results in large velocity gradients, which results in faster spinning eddies and a larger eddy diffusivity*. This process is illustrated in Figure 5.4.

The scale of the flow field is also important because *the larger eddies perform most of the transport*. One analog is moving households across the country, a common summer employment for college students. The movers would come to pack up the

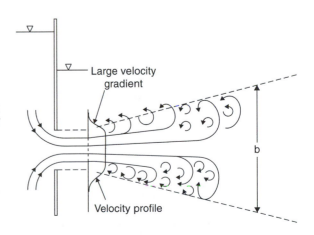

Figure 5.4. Eddy formation at the edge of a jet issuing into a tank illustrates the importance of velocity gradients in eddy diffusion coefficient.

family and load the boxes onto the semi-trailer truck. It is fair bit of work, but the family moved only from the house to the semi. These movers are like the small eddies, working hard but not getting far. When the semi is loaded, the cross-country movers transport the household belongings great distances, similar to the large eddies. The small eddies are always there in a turbulent flow, and their existence is important for local mixing. It is the large eddies, however, that are the most responsible for transport, as illustrated in Figure 5.5.

The four observations, listed previously, were enough for Ludwig Prandtl (1925) to hypothesize a simple model for describing turbulent transport that works surprisingly well, considering the complexity of turbulent flow.

C. Prandtl's Mixing Length Hypothesis for Turbulent Flow

Prandtl's mixing length hypothesis (Prandtl, 1925) was developed for momentum transport, instead of mass transport. The end result was a turbulent viscosity, instead of a turbulent diffusivity. However, because both turbulent viscosity and turbulent diffusion coefficient are properties of the flow field, they are related. Turbulent viscosity describes the transport of momentum by turbulence, and turbulent diffusivity describes the transport of mass by the same turbulence. Thus, turbulent viscosity is often related to turbulent diffusivity as

$$\varepsilon_x = \mu_{tx}/\rho \quad \varepsilon_y = \mu_{ty}/\rho \quad \text{and} \quad \varepsilon_z = \mu_{tz}/\rho \tag{5.21}$$

where μ_{tx}, μ_{ty}, and μ_{tz} are the turbulent viscosity in the x-, y-, and z-directions. Now, for the x-component of momentum (ρu), the Boussinesq approximation is

$$-\rho \,\overline{u'u'} = \mu_{tx}\frac{\partial \overline{u}}{\partial x} \tag{5.22}$$

$$-\rho \,\overline{v'u'} = \mu_{ty}\frac{\partial \overline{u}}{\partial y} \tag{5.23}$$

$$-\rho \,\overline{w'u'} = \mu_{tx}\frac{\partial \overline{u}}{\partial z} \tag{5.24}$$

Let's consider the fully developed velocity profile in the middle of a wide open channel, with x-, y-, and z-components in the longitudinal, lateral, and vertical directions, respectively. It is fully developed because $\partial \overline{u}/\partial x$ is close to zero. The fact that it is a wide channel means that $\partial \overline{u}/\partial y$ is also very small in the middle. From equations (5.22) and (5.23), we can see that the turbulent transport of momentum in the x- and

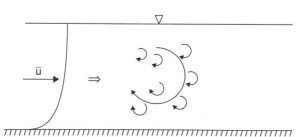

Figure 5.5. Large and small eddies in an open-channel flow. The large eddies perform most of the top to bottom transport.

y-directions will be small because the gradients are small. There will, however, be a net turbulent transport of momentum in the z-direction.

$$-\overline{w'u'} = \varepsilon_z \frac{\partial \overline{u}}{\partial z} \neq 0 \qquad (5.25)$$

Now, half of the w' values will be positive, and the other half will be negative. We will use this criteria to divide them into two parts:

$$\overline{w'u'} = \overline{w'u'}^+ + \overline{w'u'}^- \qquad (5.26)$$

where $\overline{w'u'}^+$ has a value when w' is positive and is equal to zero when w' is negative. $\overline{w'u'}^-$ has a value when w' is negative and is equal to zero when w' is positive. Consider the cases when w' is positive. Then, equation (5.26) becomes

$$\overline{w'u'} = \overline{w'u'}^+ + 0 \qquad (5.27)$$

Let us assume that an eddy of length L is pulling a blob of fluid upward, as illustrated in Figure 5.6. On average, the blob will have an x-component of velocity equal to $\overline{u}(z - L/2)$, where z is the location where u' is to be estimated. Thus, the eddy pulls up, on average, the u value that is at $z - L/2$. This will become the deviation from the temporal mean velocity at location z:

$$u' = u - \overline{u} \approx \overline{u}(z - L/2) - \overline{u}(z) \cong \frac{1}{2}\left[\overline{u}(z - L) - \overline{u}(z)\right] \qquad (5.28)$$

Equation (5.28) is a relation for a difference in velocity, which can be written as a velocity gradient times a distance:

$$u' \qquad = \qquad -\frac{\partial \overline{u}}{\partial x} \qquad \left(\frac{L}{2}\right)$$

velocity $\quad$ = $\quad$ **velocity** $\quad$ × $\quad$ **distance**
difference $\qquad\qquad$ **gradient** $\qquad\qquad$ (5.29)

Then, Equation (5.27) becomes

$$\overline{w'u'}^+ \approx \overline{\frac{w'}{2}\left[\overline{u}(z - L) - \overline{u}(z)\right]} \approx -\overline{\frac{w'}{2}}L\frac{\partial \overline{u}}{\partial z} \qquad (5.30)$$

Figure 5.6. Illustration of the relationship between velocity profile, turbulent eddies, and mixing length.

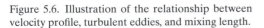

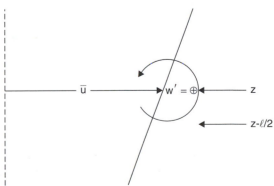

The development is similar for $\overline{w'u'}$:

$$\overline{w'u'}^- \approx \overline{w'u'}^+ \approx -\frac{\overline{w'}}{2}L\frac{\partial \overline{u}}{\partial z} \tag{5.31}$$

Now, combining equations (5.26), (5.30), and (5.31) gives

$$\overline{w'u'} = -\overline{w'L\frac{\partial \overline{u}}{\partial z}} \tag{5.32}$$

Because turbulent eddies tend to approach spherical in shape:

$$|w'| \approx |u'| \tag{5.33}$$

and from equation (5.29):

$$w' \sim L\frac{\partial u'}{\partial z} \tag{5.34}$$

If we substitute equation (5.34) into (5.32), and then substitute the result into equation (5.24), we get

$$-\overline{w'u'} = \varepsilon_z \frac{\partial \overline{u}}{\partial z} = L^2\left(\frac{\partial \overline{u}}{\partial z}\right)^2 \tag{5.35}$$

or

$$\varepsilon_z = L^2 \left|\frac{\partial \overline{u}}{\partial z}\right| \tag{5.36}$$

Equation (5.36) is *Prandtl's mixing length hypothesis*, and it works well, considering that the basis for the equation is so empirical. However, equation (5.36) does present a challenge for us: that mixing length, L, still needs to be specified. Measurements have shown us the following:

1. Near a wall, $L = \kappa z$, where κ is von Kármán's constant (von Kármán, 1930) and is very close to 0.4, and z is the distance from the closest wall.
   Prandtl also made another assumption in this region: that $w'u'$ could be approximated by a constant equal to the mean wall shear stress, or

$$-\overline{w'u'} = \tau/\rho = u_*^2 \tag{5.37}$$

Then, eliminating $\overline{w'u'}$ from equations (5.35) and (5.37) results in the well-known *logarithmic velocity profile*:

$$\frac{\overline{u}}{u_*} = \frac{1}{k}\ell n\left(\frac{z}{z_0}\right) \tag{5.38}$$

where u_* is the shear velocity at the wall, τ is the wall shear stress, and z_0 *is an* integration constant, often *called the dynamic roughness*. Table 5.2 provides some

Table 5.2: *Dynamic roughness lengths, z_0, for typical atmospheric surfaces (Turner, 1979)*

Surface type	z_0 (m)
Urban	1.3
Forest	1.3
Deciduous forest in winter	0.5
Desert shrub land	0.3
Wetland	0.3
Crop land (summer)	0.2
Crop land (winter)	0.01
Grass land (summer)	0.1
Grass land (winter)	0.001
Water	~ 0.0001

typical dynamic roughness lengths for atmospheric boundary layers. Applying equation (5.36) to (5.37) results in an equation for ε_z in this region:

$$\varepsilon_z = \kappa u_* z \qquad (5.39)$$

2. *Very* near a wall (approaching the laminar sublayer where the turbulence is so small that it is eliminated by the viscosity of the fluid); i.e., for $z u_* v < 35$, $L \sim y^2$ (Reichardt, 1951).

Making the same assumption that $u'w'$ is approximately equal to wall shear stress, this relation for L results in the following relation for velocity profile very near the wall:

$$\frac{\overline{u}}{u_*} = \beta \frac{v}{u_* z} \qquad (5.40)$$

Equation (5.39) is not used in mass transport calculations near a wall or interface because the unsteady character of mass transport in this region is very important, and equation (5.40) is for a temporal mean velocity profile. This will be discussed further in Chapter 8.

3. Away from a wall, where the closest wall does not influence the velocity profile, L is a function of another variable of the flow field (Prandtl, 1942). For example, consider the jet mixer given in Figure 5.4. In this case, the mixing length, L, is a function of the width of the jet or plume. As the jet/plume grows larger, the value of L is larger.

Here, it is easier to simply give the experimental relation for eddy diffusion coefficient:

$$\varepsilon_z = \beta\, u_{max} b \qquad (5.41)$$

where b is the width of the mixing zone, β is a constant, and u_{max} is the maximum velocity in the jet at the given location, x.

Figure 5.7 gives some relationships for eddy diffusion coefficient profiles under different conditions that will be handy in applications of turbulent diffusive transport.

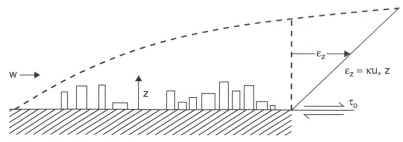

(a) Unconfined boundary layer

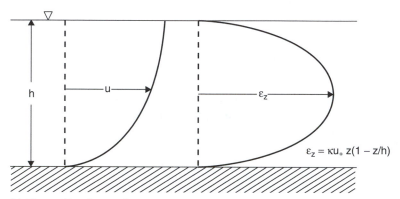

(b) Flows with a free surface

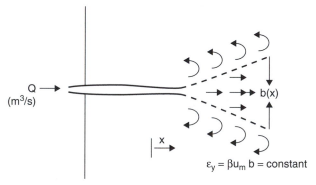

(c) Jetlike flows (not close to a boundary)

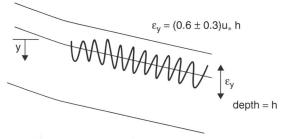

(d) Depth averaged transverse turbulent diffusion in a river

Figure 5.7. Profiles of eddy diffusion coefficient for various types of applications.

EXAMPLE 5.1: *Profile of eddy diffusion coefficient*

Estimate the eddy diffusivity profile for a wind velocity of 18 m/s measured at 10 m over a large lake (Figure E5.1.1) and calculate the elevation above the water surface where $\varepsilon_z = D$ for water vapor.

There is only one assumption needed:

The wind fetch is sufficient so that $\overline{U}_{10}$ is influenced by shear at the water surface (10 m is inside the boundary layer of the lake surface at this point).

Then, mixing length theory may be used with momentum transport to derive

$$\frac{\partial \overline{u}}{\partial z} = \frac{u_*}{\kappa z} \qquad \text{(E5.1.1)}$$

and

$$\varepsilon_z = \kappa u_* z \qquad \text{(E5.1.2)}$$

Now, Wu (1980) has provided the following equation from a fit of field data:

$$u_* = 0.01 \, \overline{u}_{10} \, (8 + 0.65 \, \overline{u}_{10})^{1/2} \qquad \text{(E5.1.3)}$$

which indicates that, as the waves get larger at high wind speeds, the boundary roughness effect on u_* increases by the factor $(8 + 0.65 \, \overline{u}_{10})^{1/2}$, where $\overline{u}_{10}$ is given in m/s.

Then,

$$\varepsilon_z(\text{m}^2/\text{s}) = 0.01 \, \kappa \, \overline{u}_{10} z \, (8 + 0.65 \, \overline{u}_{10})^{1/2} = 0.32 \, z \qquad \text{(E5.1.4)}$$

when z is given in meters. Now, the diffusion coefficient of water vapor in air was calculated in Example 3.3 to be $D = 2.6 \times 10^{-5} \, \text{m}^2/\text{s}$. Then, the elevation at which the diffusivity of water vapor would equal eddy diffusivity in this case would be

$$0.32 \, z = 2.6 \times 10^{-5} \qquad \text{(E5.1.5)}$$

or

$$z = 8 \times 10^{-5} \text{m} = 0.08 \text{ mm} = 80 \, \mu \qquad \text{(E5.1.6)}$$

In Example 5.1, equation (E5.1.6) gives $z = 80 \, \mu$ elevation above the water surface to have eddy diffusivity equal to the diffusivity of water. A similarly small elevation

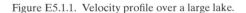

Figure E5.1.1. Velocity profile over a large lake.

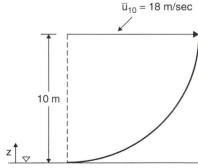

would result for almost any environmentally relevant compound. We can thus see that both ε and D need to be considered simultaneously in equation (5.20) *only* very close to surfaces in turbulent flow, where ε approaches the diffusion coefficient. Otherwise, diffusivities can be ignored in solving turbulent flow transport problems, because $\varepsilon + D$ is essentially equal to ε.

EXAMPLE 5.2: *Concentration profile of suspended sediment in a river (assuming ε_z is constant)*

We will apply equation (5.20) to solve for the concentration profile of suspended sediment in a river, with some simplifying assumptions. Suspended sediment is generally considered similar to a solute, in that it is a scalar quantity in equation (5.20), except that it has a settling velocity. We will also change our notation, in that the bars over the temporal mean values will be dropped. This is a common protocol in turbulent transport and will be followed here for conformity. Thus, if an eddy diffusion coefficient, ε, is in the transport equation,

$$
\begin{array}{lll}
u & \text{means} & \bar{u} \\
v & \text{means} & \bar{v} \\
w & \text{means} & \bar{w} \\
C & \text{means} & \bar{C}
\end{array}
$$

throughout the remainder of this text. This may make differentiation between the instantaneous value of u, v, w, and C and the temporal mean value difficult, but it is easier with this convention to write the diffusion equation on the chalkboard or in one's notes. Figure E5.2.1 gives a longitudinal and lateral cross section of our river. We will make the following assumptions:

1. The flow is steady over the long term, so that $\partial C/\partial t = 0$.
2. The flow is fully developed, such that any gradient with respect to x is equal to zero ($\partial C/\partial x = 0$).
3. The river can be divided into a series of longitudinal planes with no significant interaction, such that $v = 0$ and $\varepsilon_y = 0$. (This is the assumption of the stream-tube computational models.)
4. The vertical eddy diffusivity, ε_z, is a constant value.

Assumption 3 and 4 are the more difficult to justify.

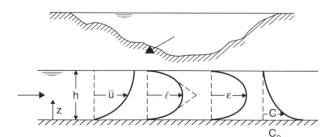

Figure E5.2.1. Lateral and longitudinal cross-sections of a typical river.

The solute will have a vertical velocity, $w = -v_s$, where v_s is the settling velocity of the suspended sediment. Then, equation (5.20) becomes

$$-v_s \frac{\partial C}{\partial z} = \frac{\partial}{\partial z}\left[(D + \varepsilon_z)\frac{\partial C}{\partial z}\right] \tag{E5.2.1}$$

where we have not yet applied assumption 4. We can move the settling velocity into the partial term:

$$\frac{\partial(-v_s C)}{\partial z} = \frac{\partial}{\partial z}\left[(D + \varepsilon_z)\frac{\partial C}{\partial z}\right] \tag{E5.2.2}$$

and since both sides of equation (E5.2.2) are a gradient with respect to z, the terms inside of the gradients must also be equal:

$$-v_s C = (D + \varepsilon_z)\frac{dC}{dz} \tag{E5.2.3}$$

Equation (E5.2.3) is converted to an ordinary differential equation because all variables are only a function of z. Now, we will deal with assumption 4. Figure 5.7 gives the equation developed by Rouse (1937) for ε_z:

$$\varepsilon_z = \kappa u_* z (1 - z/h) \tag{E5.2.4}$$

where u_* is the shear velocity at the bottom of the channel, or

$$u_* = \sqrt{\tau/\rho} \tag{E5.2.5}$$

where τ is the shear stress at the wall. For a fully developed open channel flow in a wide channel, the following relation is derived from the energy equation (Nezu and Nakagawa, 1993):

$$u_* = \sqrt{gh S} \tag{E5.2.6}$$

This derivation can be found in a text on fluid mechanics or open channel flow. Assumption 4 states that $\varepsilon_z = \overline{\varepsilon}_z$ for all values of z, where ε_z is the depth average, or

$$\overline{\varepsilon}_z = \frac{1}{h}\int_0^h \varepsilon_z dz = \frac{\kappa u_*}{h}\int_o^h z(1 - z/h)dz = 0.067\, u_* h \tag{E5.2.7}$$

where h is the depth of the stream. The term $\overline{\varepsilon}_z$ is almost always much greater than D in a turbulent flow. Thus,

$$D + \overline{\varepsilon_z} \cong \overline{\varepsilon_z} \tag{E5.2.8}$$

Now, substituting equation (E5.2.8) into (E5.2.7), then equation (E5.2.7) into (E5.2.8), and finally equation (E5.2.8) into (E5.2.3) results in

$$\overline{\varepsilon_z}\frac{dC}{dz} + v_s C = 0 \tag{E5.2.9}$$

We will solve equation (E5.2.9) by separating variables,

$$\frac{dC}{C} = \frac{-v_s}{\overline{\varepsilon}_z}dz \tag{E5.2.10}$$

integrating and taking both sides of the solution to the power of e:

$$C = \beta_1\, e^{\frac{-v_s}{\overline{\varepsilon}_z}z} \tag{E5.2.11}$$

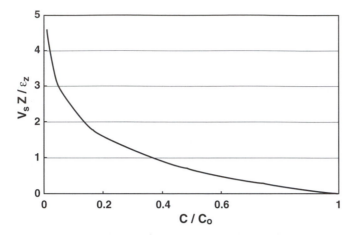

Figure E5.2.2. Suspended sediment concentration profile for Example 5.2.

Now, we need a boundary condition to determine β_1. This is difficult with suspended sediment profiles. We can develop a fairly good estimate of the distribution of suspended sediment once we have a known concentration at some location in the flow field. In the sediment transport field, bed load and suspended load are often discussed. The relation between the two, and some experience and measurements of both simultaneously, can be used to predict an equivalent suspended sediment concentration at the bed. Then, the relevant boundary condition of equation (E5.2.11) is

$$\text{At } z = 0, \ C = C_0.$$

where C_0 is the concentration that has been determined from the bed load-suspended load relationship. Applying this boundary condition to equation (E5.2.11) gives $\beta_1 = C_0$, and our solution is

$$C = C_0 \, e^{-v_s z / \bar{\varepsilon}_z} \tag{E5.2.12}$$

The result is illustrated in Figure E5.2.2. This problem can also be solved *without* assumption 4. The solution will be left to the reader.

We will now turn our attention from one-dimensional solutions of the diffusion equation to two-dimensional solutions.

EXAMPLE 5.3: *River mixing zone for waste discharge (evaluated as a point source with transverse diffusion)*

The discharged effluent from an agricultural wastewater treatment lagoon enters the Minnesota River with some residual pollutant concentrations. When completely mixed into the river, these pollutants will meet concentration criteria put forward by the state agencies. There is, however, a region before the waste stream approaches a perfectly mixed condition, called a mixing zone, where the state criteria may not

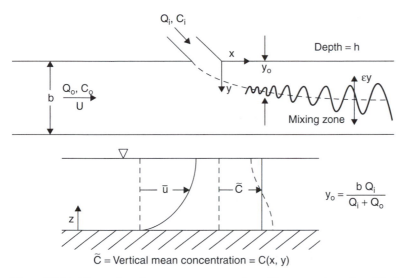

Figure E5.3.1. Illustration of a sewage treatment plant effluent entering a river. (*Top*) Plan view. (*Bottom*) Longitudinal section illustrating $\tilde{C}$.

be met. Develop an equation that can be used to determine the concentration of compounds in this criteria-exceedence region for the various pollutants of the waste water effluent.

If the effluent discharge is low, compared with the river discharge, we can approach this effluent as a point source, handled as a Dirac delta. As a first approximation, this is not a bad approach. In fact, we will utilize the solution from Example 2.2.

The river and waste effluent is illustrated in Figure E5.3.1. The river has a discharge of Q_0, a cross-sectional mean velocity of U, a mean stream width of b, and an ambient pollutant concentration of C_0. The effluent discharge is Q_i, with a pollutant concentration of C_i. We will solve for a vertical mean concentration, $\tilde{C}$, which varies spatially across the river (y-axis) and in the direction of flow (x-axis), or

$$\tilde{C} = \frac{1}{h} \int_0^h C \, dz \qquad (E5.3.1)$$

where h is river depth.

We will also make the following assumptions:

1. Steady state, $\partial \tilde{C} / \partial t = 0$.
2. No reactions of the compound of interest, and no loss to the air or sediments, $S = 0$. (This is a conservative assumption, if one is interested in pollutant concentrations.)
3. Temporal mean velocities in the vertical and lateral directions are zero, $v = w = 0$.
4. Lateral gradients are much greater than longitudinal gradients, $\partial \tilde{C} / \partial y \gg \partial \tilde{C} / \partial x$.
5. The mean lateral eddy diffusion coefficient is much greater than the diffusion coefficient for any compound, $\overline{\varepsilon}_y \gg D$.

Then, equation (5.20) becomes

$$U\frac{\partial \tilde{C}}{\partial x} = \overline{\varepsilon}_y \frac{\partial^2 \tilde{C}}{\partial y^2} \tag{E5.3.2}$$

We are using an equivalent vertical-mean eddy diffusion coefficient because we are computing a vertical mean concentration.

The boundary conditions for this example are:

1. As we move downstream, the concentration approaches one, which is well mixed, or $C \Rightarrow C_m$ as $x \Rightarrow \infty$, where

$$C_m = \frac{Q_0 C_0 + Q_i C_i}{Q_0 + Q_i} \tag{E5.3.3}$$

2. No net flux of compound at $y = b$
3. Continuous point source (representing the effluent) with a given mass flux rate at $(x, y) = (0,0)$

Our solution technique will be to adapt the governing equation and the boundary conditions from Example 2.2 to apply to this problem. We spent considerable effort proving to ourselves that equation (E2.2.3) was a solution to the governing differential equation. We can avoid that tedious process by showing that this problem has a similar solution, except with different variables and different constants.

First, let us consider equation (E2.2.2) and compare it with equation (E5.3.2):

$$\frac{\partial C}{\partial t} = D\frac{\partial^2 C}{\partial z^2} \tag{E2.2.2}$$

The primary difference between the two equations is the unsteady term in equation (E2.2.2) and the convective term in equation (E5.3.2). Now, let's convert our coordinate system of Example 2.2 to a moving coordinate system, moving at the bulk velocity, U, which suddenly experiences a pulse in concentration as it moves downstream. This is likened to assuming that we are sitting in a boat, moving at a velocity U, with a concentration measuring device in the water. The measurements would be changing with time, as we moved downstream with the flow, and the pulse in concentration would occur at $x = 0$. We can therefore convert our variables and boundary conditions as follows:

Equation (E5.2.2)		Equation (E5.3.2)
t	$\Rightarrow$	x/U
z	$\Rightarrow$	y
D	$\Rightarrow$	ε_y
Boundary conditions:		
Dirac delta at $(z,t) = (0, 0)$	$\Rightarrow$	Dirac delta at $(x, y) = (0, 0)$
$C \Rightarrow 0$ as $t \Rightarrow \infty$	$\Rightarrow$	$\tilde{C} \Rightarrow 0$ as $x \Rightarrow \infty$
Background concentration of 0	$\Rightarrow$	Background concentration of C_0

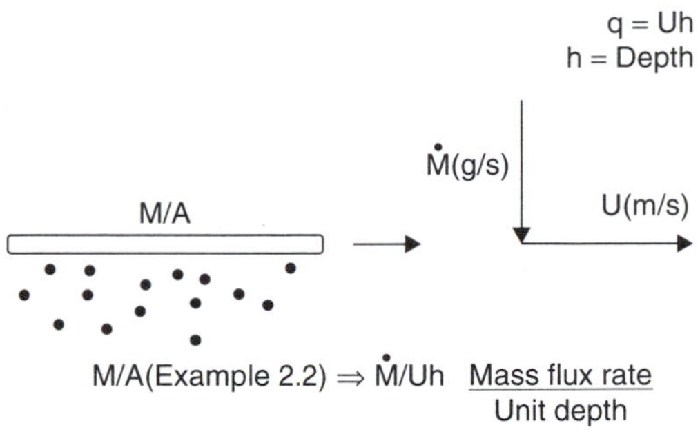

Figure E5.3.2. Adjustment of the boundary conditions from a static pulse to a continuous pulse with mass flux rate, M.

The fully mixed concentration can be taken care of by placing an image source in the solution at $(x,y) = (0, 2b)$. This will meet the conditions of boundary conditions 1 and 2. The ambient concentration will be handled by assigning $\hat{C} = \tilde{C} - C_0$. Then, equation (E5.3.2) becomes

$$U\frac{\partial \hat{C}}{\partial x} = \bar{\varepsilon}_y \frac{\partial^2 \hat{C}}{\partial y^2} \tag{E5.3.4}$$

Equation (E5.3.4) has the same form as equation (E5.3.2) because the constant C_0 drops out of an equation consisting of only differential terms. Then, at negative values of x, $C = C_0$ and $\hat{C} = 0$, similar to the boundary conditions for Example 2.2 at negative values of time.

We also need to make sure that the strength of the Dirac delta has a similarity between the two solutions. In Example 2.2, the strength was a mass per unit area, or M/A. For this problem, because we are following a moving control volume, we need a mass flux rate, $\dot{M}$, and the unit area passing the outfall per unit time (an "area flux rate"), Uh. Thus, similarity between the two solutions means that $A \Rightarrow Uh$ and $M/A \Rightarrow \dot{M}/(Uh)$, as illustrated in Figure (E5.3.2).

Combining these variable transformations with the addition of an image solution results in the solution to equation (E5.3.2):

$$\hat{C} = \frac{2\dot{M}}{h(4\pi \bar{\varepsilon}_y xU)^{1/2}} \left\{ \exp\left(\frac{-Uy^2}{4\bar{\varepsilon}_y x}\right) + \left[\frac{-U(y - 2b)^2}{4\bar{\varepsilon}_y x}\right] \right\} \tag{E5.3.5}$$

Once we know our mass flux rate, we can estimate the concentrations in the river with our analytical solution. The incremental mass flux into the stream due to the effluent is based on a mass balance at the outfall and given by the equation

$$\frac{\dot{M}}{Uh} = \frac{(C_i - C_0)Q_i}{Uh} = \frac{\hat{C}_i Q_i}{Uh} \tag{E5.3.6}$$

Then, equation (E5.3.5) becomes

$$\frac{\hat{C}}{\hat{C}_i} = \frac{Q_i}{h(\pi \bar{\varepsilon}_y xU)^{1/2}} \left\{ \exp\left(\frac{-Uy^2}{4\bar{\varepsilon}_y x}\right) + \left[\frac{-U(y - 2b)^2}{4\bar{\varepsilon}_y x}\right] \right\} \tag{E5.3.7}$$

The solution of equation (E5.3.7) is accurate when we are not too close to the Dirac delta source at $(x,y) = (0,0)$. Of course, we cannot say how close is too close unless we compare it with a more accurate solution. The next example will provide an equation to describe a solution that does not have a source term of infinitely small width and infinite concentration. This solution can be compared with equation (E5.3.7) to determine the answer to the "how close is too close" question.

EXAMPLE 5.4: *The mixing of two streams (concentration front with moving coordinates and first order and zero-order source terms)*

The setup for this problem will be similar to Example 5.3, except that the effluent discharge, Q_i, is *not* small, compared with river discharge, Q. The solution could also be used for the mixing of two rivers that have differing concentrations of the relevant compounds, as illustrated in Figure E5.4.1. In this figure, y_0 is the width of the stream taken by the effluent discharge, or

$$y_0 = \frac{b\, Q_i}{Q_i + Q_0} \tag{E5.4.1}$$

With the first-order and zero-order source terms, the governing equation becomes

$$U\frac{\partial \tilde{C}}{\partial x} = \bar{\varepsilon}_y \frac{\partial^2 \tilde{C}}{\partial y^2} - k(C - C_E) \tag{E5.4.2}$$

where C_E is the equilibrium value of C, and the other variable definitions are the same as Example 5.3. Stefan and Gulliver (1978) have developed a solution that can be applied to this problem:

$$\frac{\hat{C}}{\hat{C}_i} = \frac{1}{2}\left(\operatorname{erf} R_1 + \operatorname{erf} R_3 - \operatorname{erf} R_2 - \operatorname{erf} R_4\right)e^{-kx/U} + \frac{\hat{C}_E}{\hat{C}_i}\left(1 - e^{-kx/U}\right) \tag{E5.4.3}$$

where

$$\hat{C}_E = C_E - C_0$$

$$R_1 = \frac{y + y_0}{\sqrt{4x\,\bar{\varepsilon}_y/U}}$$

$$R_2 = \frac{y - y_0}{\sqrt{4x\,\bar{\varepsilon}_y/U}}$$

$$R_3 = \frac{y + y_0 - 2b}{\sqrt{4x\,\bar{\varepsilon}_y/U}}$$

and

$$R_4 = \frac{y - y_0 - 2b}{\sqrt{4x\,\bar{\varepsilon}_y/U}}$$

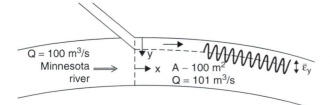

Figure E5.4.1. Illustration of the mixing zone at the junction of two rivers.

R_1 and R_2 are concentration fronts that occur at $y = -y_0$ and $y = y_0$, respectively. They form a front and its image about the y-axis so that there is no flux of mass across the $y = 0$ plane. R_3 and R_4 are images of the R_1 and R_2 concentration fronts such that there is no mass flux across the $y = b$ plane. Of course, there is now a lack of symmetry about the $y = 0$ plane, and if we go far enough downstream, we will find leakage across this plane because the R_3 and R_4 concentration fronts have influenced concentration profiles at $y = 0$. We could solve this problem more precisely by adding more images to our solution (e.g., at $y = -2b - y_0$ and at $y = -2b + y_0$), but an infinite number of images would be needed for an exact solution. In addition, the river would probably have changed significantly by the time we get far enough downstream, so the solution would no longer apply.

Similar concerns involving turbulent transport of chemicals occur in the air environment, except it is difficult to confine the problem to two dimensions. Example 5.5 is a demonstration of the application of equation (5.20) to environmental transport in the air.

EXAMPLE 5.5: *Concentration of organic compounds released into the air by an industrial plant (application of the product rule to error function solutions)*

There is some concern about the emissions from the adhesives produced in an industrial plant. Specifically, the town of Scream Hollow is 1 km away from the plant, where citizens have begun to complain about odors from the plant and headaches. One culprit, aside from a haunting, may be the release of acrolein, C_3H_4O, a priority pollutant that is an intermediary of many organic reactions. The average release from the 200 m $\times$ 200 m $\times$ 10 m plant is assumed to be 20 g/hr. If the wind is blowing directly toward Scream Hollow, at 3 m/s measured at 3 m height, with a dynamic roughness of 0.2 m for the farmland, what concentrations will the Scream Hollow inhabitants experience? Is this above the EPA (Environmental Protection Agency) threshold limit of 0.1 ppm(v)?

We will need to make some assumptions to formulate this problem. They are:

1. The acrolein release is distributed over the most downwind plane of the building. With the important concentration being 1 km away, this is not a bad assumption. Then, the acrolein will be released over the plane that is 200 m $\times$ 10 m. If 20 g/hr = 0.0056 g/s are released into a wind moving at 3 m/s, the initial concentration is

$$C_0 = \frac{0.0056 \text{ g/s}}{3 \text{ m/s}(200 \text{ m})(10 \text{ m})} = 9.3 \times 10^{-7} \text{ g/m}^3 \qquad \text{(E5.5.1)}$$

Figure E5.5.1. Illustration of toxic chemical release into the atmosphere, with the wind blowing toward a town.

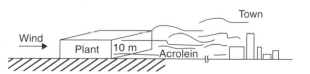

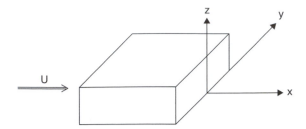

Figure E5.5.2. Illustration of the coordinate system for Example 5.5.

2. We will use a cross-sectional mean velocity of $U = \bar{u}$ at 3 m height or $U = 3\,\mathrm{m/s}$.
3. We will use $\bar{\varepsilon}_z = \bar{\varepsilon}_y = \varepsilon_z$ at 3 m height.
4. We will not consider any of the source or sink terms for acrolein.

We will also set up the coordinates so that $(x, y, z) = (0,0,0)$ occurs on the ground at mid-plant width and will orient the wind in the x-direction.

With these assumptions, the governing equation becomes:

$$U\frac{\partial C}{\partial x} = \bar{\varepsilon}_y \frac{\partial^2 C}{\partial y^2} + \bar{\varepsilon}_z \frac{\partial^2 C}{\partial z^2} \tag{E5.5.2}$$

The boundary conditions are:

1. At $(x, y, z) = (0, -100\,\mathrm{m} \Rightarrow 100\,\mathrm{m}, 0 \Rightarrow 10\,\mathrm{m})$, $C = C_0$
2. As $x \Rightarrow \infty$, $y \Rightarrow \infty$, or $z \Rightarrow \infty$, $C \Rightarrow 0$
3. Zero mass flux at $z = 0$

These boundary conditions, illustrated in Figure E5.5.2, will give us a concentration front, but in two dimensions. In addition, we have a zero flux condition that will require an image solution. We will use the solution of Example 2.7 to develop a solution for this problem. The solution, before applying boundary conditions, was

$$C = \beta_0 + \beta_1 \mathrm{erf}\left(\frac{z}{\sqrt{4Dt}}\right) \tag{E2.7.5}$$

Now, we need an image to the concentration front about the $z = 0$ plane. In the y-direction, we have a step up at $y = -\Delta y$ and a step down at $y = \Delta y$. We will also use the product rule (Example 2.3) to indicate that the solution to our governing equation (E5.5.2) for the y-direction should be multiplied times the solution in the z-direction. Then, the solution can be given as

$$\frac{C}{C_0} = \beta_0 + \left\{ \beta_1 \mathrm{erf}\left[\frac{(z+\Delta z)}{\sqrt{4\bar{\varepsilon}_z x/U}}\right] + \beta_2 \mathrm{erf}\left[\frac{(z-\Delta z)}{\sqrt{4\bar{\varepsilon}_z x/U}}\right] \right\}$$
$$\times \left\{ \beta_3 \mathrm{erf}\left[\frac{(y+\Delta y)}{\sqrt{4\bar{\varepsilon}_y x/U}}\right] + \beta_4 \mathrm{erf}\left[\frac{(y-\Delta y)}{\sqrt{4\bar{\varepsilon}_y x/U}}\right] \right\} \tag{E5.5.3}$$

where $\Delta y = 100\,\mathrm{m}$ and $\Delta z = 10\,\mathrm{m}$. If we plug equation (E5.5.3) into equation (E5.5.2), we will find that equation (E5.5.3) is a solution to our governing equation.

Now, to see if our boundary conditions can be satisfied with the form of equation (E5.5.3):

1. At $x \Rightarrow \infty$, $C \Rightarrow 0$. Thus, $\beta_0 = 0$.
2. At $z \Rightarrow \infty$, $C \Rightarrow 0$. Thus, $\beta_1 = -\beta_2$.
3. At $y \Rightarrow \infty$, $C \Rightarrow 0$. Thus, $\beta_3 = -\beta_4$.
4. At $x \Rightarrow 0$, and $(y, z) = (0,0)$, $C/C_0 = 1$.

With the last boundary condition, equation (E5.5.3) becomes

$$1 = (\beta_1 - \beta_2)(\beta_3 - \beta_4) \tag{E5.5.4}$$

or

$$1 = 2\beta_1 \times 2\beta_3 \tag{E5.5.5}$$

or

$$1 = 2\beta_2 \times 2\beta_4 \tag{E5.5.6}$$

Finally, at $x \Rightarrow 0$ and $(y, z) = (0, \Delta z)$, $C/C_0 = \frac{1}{2}$. Thus, $\beta_1 = \frac{1}{2}$.

Applying this last boundary condition to equations (E5.5.5) and (E5.5.6) results in $\beta_3 = \frac{1}{2}$, $\beta_2 = -\frac{1}{2}$, and $\beta_4 = -\frac{1}{2}$. Thus, the solution to equation (E5.5.2) is

$$
\begin{aligned}
\frac{C}{C_0} = \frac{1}{2} &\left\{ \mathrm{erf}\left[\frac{(z + \Delta z)}{\sqrt{4\bar{\varepsilon}_z x / U}} \right] - \mathrm{erf}\left[\frac{(z - \Delta z)}{\sqrt{4\bar{\varepsilon}_z x / U}} \right] \right\} \\
\times &\left\{ \mathrm{erf}\left[\frac{(y + \Delta y)}{\sqrt{4\bar{\varepsilon}_y x / U}} \right] - \mathrm{erf}\left[\frac{(y - \Delta y)}{\sqrt{4\bar{\varepsilon}_y x / U}} \right] \right\}
\end{aligned}
\tag{E5.5.7}
$$

Now, if we use $\Delta z = 10 \, \mathrm{m}$, $\Delta y = 100 \, \mathrm{m}$, $U = 3 \, \mathrm{m/s}$, the only remaining parameter to find is $\bar{\varepsilon}$. Using the equation given in Example 5.1:

$$\bar{\varepsilon}_z = \bar{\varepsilon}_y = \kappa u_* z \tag{E5.1.1}$$

Note that the logarithmic boundary profile can be written as

$$\frac{\bar{u}}{u_*} = \ln\left(\frac{z}{z_0} \right) \tag{E5.5.8}$$

where z_0 is the dynamic roughness, assumed to be 0.2 m for the crop land between the plant and Scream Hollow. Then,

$$u_* = \frac{\bar{u}}{\ln(z/z_0)} = \frac{3 \, \mathrm{m/s}}{\ln\left(\dfrac{3 \, \mathrm{m}}{0.2 \, \mathrm{m}} \right)} = 1.1 \, \mathrm{m/s} \tag{E5.5.9}$$

and

$$\bar{\varepsilon}_z = 0.4 \, (1.1 \, \mathrm{m/s})(3 \, \mathrm{m}) = 1.3 \, \mathrm{m}^2/\mathrm{s} \tag{E5.5.10}$$

If we now plug all of the parameters for the industrial plant into equation (E5.5.7), we get $C = 0.25 \, \mu\mathrm{g/m}^3 = 2.5 \times 10^{-7} \mathrm{g/m}^3$. In terms of ppm(v), we will use

$\rho_{air} = 1.2\,g/m^3$, and the molecular weights of air and acrolein of 29 and 56 g/mole, respectively. Then,

$$C = \frac{2.5 \times 10^{-7}\,g/m^3}{\rho_{air}}\frac{MW_{air}}{MW_{C_3H_4O}} = \frac{2.5 \times 10^{-7}\,g/m^3}{1.2\,g/m^3}\frac{20\,g/mole}{56\,g/mole}$$ (E5.5.11)

$$= 1.08 \times 10^{-7}\frac{moles\ C_3H_4O}{mole\ air}$$

This is right at the threshold for continuous exposure, and the pollution from the plant should be investigated in more detail.

D. Problems

1. A styrene spill has been discovered in a large lake. Right on the water surface ($z = 0$), the styrene concentration in the air over the lake is 4 µg/m³. Assuming that $[C_8H_8]_{z=0}$ stays constant and $[C_8H_8]_{z=1,000} = 1\,ng/m^3$, determine and plot the air concentration profile at wind speeds (measured at 10 m) of 0.1 m/s, 1 m/s, and 10 m/s.

2. The mean suspended sediment concentration of a 3-m-deep river is $10\,g/m^3$. The mean sediment settling velocity is 0.01 m/s, and the river slope is 2×10^{-3}. Assuming that ε_z is constant with z, what is the concentration profile for sediment in the river? Does the solution make physical sense at all boundaries? Explain.

3. Derive the concentration profile for problem 2, except with ε_z as a function of z. Plot the concentration profile on the same plot. What accounts for the difference? *Hint*: Set $C = C_0$ at $z = 1$ mm, then determine C_0 value that results in the given mean suspended concentration.

4. A 0.5 mg/m³ concentration of total polychlorinated biphenols are discharged at 1 m³/s. from a St. Paul manufacturing plant into the Mississippi River. Assuming the problem can be modeled as a point source, plot the concentration isopleths versus horizontal distance downstream of the plant, with no reaction rates, to the end of the mixing zone. River data: $Q = 20$ m³/s; mean width $= 200$ m, mean depth $= 2$ m; slope $= 10^{-4}$.

5. You are interested in the near-field mixing of the total polychlorinated biphenols in problem 4, where the point source does not apply. Estimate the near-field concentrations of total polychlorinated biphenols, plot the isopleths of concentration and compare the results with those from your point source solution.

6 Reactor Mixing Assumptions

Solving the diffusion equation in environmental transport can be challenging because only specific boundary conditions result in an analytical solution. We may want to consider our system of interest as a reactor, with clearly defined mixing, which is more amenable to time dependent boundary conditions. The ability to do this depends on how well the conditions of the system match the assumptions of reactor mixing. In addition, the system is typically assumed as one dimensional. The common reactor mixing assumptions are as follows:

Complete Mix Reactor – The complete mix reactor is also labeled a completely stirred tank reactor. It is a container that has an infinite diffusion coefficient, such that any chemical that enters the reactor is immediately mixed in with the solvent. In Example 2.8, we used the complete mix reactor assumption to estimate the concentration of three atmospheric pollutants that resulted from an oil spill. We will use a complete mix reactor (in this chapter) to simulate the development of high salt content in dead-end lakes. A series of complete mix reactors may be placed in series to simulate the overall mixing of a one-dimensional system, such as a river. In fact, most computational transport models *are* a series of complete mix reactors.

Plug Flow Reactor – A plug flow reactor is a one-dimensional reactor with no mixing and a uniform velocity profile. In other words, the diffusion coefficient is equal to zero. A tracer pulse placed into a truly plug flow reactor would leave the reactor, later, with the same concentration profile that was put into the inflow. We will show how, in a historic paper by Streeter and Phelps (1925), plug flow was used to simulate the oxygen sag in a river below a waste effluent that had a high biochemical oxygen demand. An environment without mixing, however, is not often realistic, and plug flow does not provide an appropriate model for many environmental transport situations.

Dead Zones – Dead zones in a complete mix reactor do not participate in the mixing process. They effectively reduce the reactor size, with no exchange between the dead zone and the reactor. An example might be a wetland at the edge of

Table 6.1: *Systems that are most often simulated by complete mix reactors, plug flow reactors, etc.*

Complete mix reactors	Plug flow or complete mix reactors in series
Lakes	Rivers
Estuaries	Groundwater
Dechlorination chambers	Conduit flow
Chlorination chambers	Filters
Activated sludge	Air strippers

a river, which only exchanges water with the river at high flows. At low flows, it could be considered a dead zone. A dead zone with some exchange between the main reactor and the dead zone can also be used for environmental modeling. These leaky dead zones can be used to simulate a wetland, an embayment, a harbor, and so on, in a more realistic fashion.

Bypass – A bypass is a parallel path taken by a portion of the fluid that skips the reactor altogether without any residence time. As with dead zones, a bypass is not very useful unless a plug flow or mixed tank reactor is placed into the bypass, resulting in a parallel-flow reactor system.

Feedback – Feedback describes the removal of a portion of the fluid from the tail end or the middle of a system, moving it back toward the inflow. Feedback is often used in reactors that utilize bacterially mediated reactions to ensure a bacterial population. For example, activated sludge systems that are oriented in a channel-type of arrangement often need feedback to perform properly.

Plug Flow with Dispersion – Plug flow with dispersion is a concept that is often used to describe one-dimensional flow systems. It is somewhat more flexible in computational models because the mixing within the system is not dependent on reactor size, as with complete mix tanks in series. Plug flow with dispersion will be described in the second half of this chapter because special techniques are needed for the analysis.

The various systems that are modeled by complete mix reactors and plug flow reactors are given in Table 6.1.

A. Complete Mix Reactors

A complete mix reactor is one with a high level of turbulence, such that the fluid is immediately and completely mixed into the reactor. The outflow concentration and the reactor concentration are equal, and the diffusion term is zero due to the gradient being zero. Figure 6.1 shows an illustration of the concept. If we make the entire reactor into our control volume, then a mass balance on the reactor gives

$$\text{Rate of Accumulation} = \text{Flux rate IN} - \text{Flux rate OUT} + \text{Source} - \text{sink rates} \tag{6.1}$$

$$\mathcal{V}\frac{dC}{dt} = QC_i - QC + \mathcal{V}S \tag{6.2}$$

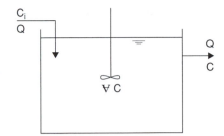

Figure 6.1. Illustration of a complete mix reactor.

where Q is the discharge into and out of the reactor, and C_i is the inflow to the reactor, which may vary with time. Equation (6.2) may be rearranged to give us

$$\frac{dC}{dt} = \frac{1}{t_r}(C_i - C) + S \tag{6.3}$$

where t_r is the residence time of the reactor, or $\forall/Q$.

If the inflow and outflow are zero, our complete mix reactor becomes a *batch reactor*, as utilized in Example 6.1 to relate half-life to a first-order rate constant.

EXAMPLE 6.1: *Compound half-life*

Determine the half-life of vinyl chloride in air with a given concentration of carbon monoxide and nitrous oxides (the main components of smog).

The half-life of vinyl chloride, C_2H_3Cl, can be determined using a batch reactor. The components of reaction and the reactor are given in Figure E6.1.1. Without an inflow or outflow and with first-order kinetics, assuming that the other components in the reaction have a large concentration, equation (6.3) becomes

$$\frac{dC}{dt} = -kC \tag{E6.1.1}$$

with boundary conditions

$$\text{at } t = 0, C = C_0$$

Figure E6.1.1. Batch reactor used to determine the half-life of vinyl chloride in smog.

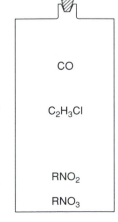

Now, separating variables and integrating equation (E6.1.1) gives

$$\int_{C_0}^{C} \frac{dC^*}{C^*} = \int_{0}^{t} -k\,dt^* \tag{E6.1.2}$$

which results in

$$\ln C - \ln C_0 = -kt \tag{E6.1.3}$$

or

$$\frac{C}{C_0} = e^{-kt} \tag{E6.1.4}$$

where the superscript "*" indicates dummy variables of integration. Data required are the concentration of vinyl chloride, carbon monoxide, and nitrous oxides in the batch reactor and the time that the concentration occurred. Then, a plot of log C versus time will be fit with a least-squares regression with a slope of $-k$.

The half-life, $t_{1/2}$ is the time required to teach $C_t/C_0 = 1/2$, or

$$\frac{C_t}{C_0} = \frac{1}{2} = e^{-kt_{1/2}} \tag{E6.1.5}$$

Thus,

$$t_{1/2} = \frac{\ln_{1/2}}{k} = \frac{0.69}{k} \tag{E6.1.6}$$

Batch reactors are traditionally used to determine a half-life of a chemical.

A complete mix reactor can also be used to simulate a larger water body when the process under consideration has a larger time scale than the mixing processes in the lake, as the next example demonstrates.

EXAMPLE 6.2: *Saline lake formation*

The Great Salt Lake in Utah has a salinity that is increasing over time. It is said that the salty character of these waters occurs because of evaporation of a dead-end lake that has no outflow. How does this happen?

The process is illustrated in Figure E6.2.1. In the case of the Great Salt Lake, the volume, $\forall$, is $5 \times 10^{11}\ \mathrm{m^3}$ and the discharge into the lake, Q, is $3 \times 10^9\ \mathrm{m^3/yr}$. In addition, the inflow concentration of sodium, C_i, is 0.2 g/L. Beginning at an in-lake concentration of 0.2 g/L, we will determine how long it will take to reach the current concentration of 100 g/L.

Figure E6.2.1. Illustration of the increase in
salinity over time in the Great Salt Lake.

We will make two assumptions:

1. *The Great Salt Lake can be considered as a complete mix reactor for this problem.*
   This is a valid assumption because the time scale of the salinity changes that lead
   to a salty lake is on the order of tens of thousands of years.
2. *The current conditions regarding residence time and evaporation rate apply to the
   past.* With no reaction utilizing sodium in the lake, the mass transport equation
   for the complete mixed reactor is

$$\forall \frac{dC}{dt} \qquad = \qquad QC_i \qquad - \qquad 0$$

Rate of $\quad$ = **Rate of flux** − **Rate of flux**

Accumulation $\qquad\qquad$ **IN** $\qquad\qquad$ **OUT**

$(E6.2.1)$

Integrating from time $t = 0$:

$$\int_{C_0}^{C} dC^* = \frac{Q}{\forall} C_i \int_{0}^{t} dt^* \qquad\qquad (E6.2.2)$$

or

$$C = C_i + \frac{Q}{\forall} C_0 t \qquad\qquad (E6.2.3)$$

The results of equation (E6.2.3), given in Table E6.2.1, indicate that, under the
current conditions, it would have taken roughly 85,000 years to reach the current
salinity of the Great Salt Lake. This, of course, ignores the fact that the Great
Salt Lake has shrunk considerably over the years from Lake Bonneville, leaving
the Bonneville Salt Flats, and that past flows into the lake were not at the current
value. The actual age of the lake is 18,000 to 25,000 years; so, our assumptions
were relatively accurate (i.e., the correct order of magnitude).

Table E6.2.1: *Development of
sodium concentrations in the
Great Salt Lake over time*

t (yr)	C (g/L)
10	0.212
1,000	1.4
10,000	12.2
50,000	60.1
83,167	100
100,000	120

Figure 6.2. Illustration of a plug flow reactor.

B. Plug Flow Reactor

In the ideal plug flow reactor, the flow traverses through the reactor like a plug, with a uniform velocity profile and no diffusion in the longitudinal direction, as illustrated in Figure 6.2. A nonreactive tracer would travel through the reactor and leave with the same concentration versus time curve, except later. The mass transport equation is

$$\frac{\partial C}{\partial t} d\mathcal{V} \qquad = U\left(C - C - \frac{\partial C}{\partial x} dx\right) dA + S d\mathcal{V}$$

Rate of $\quad$ = $\quad$ **Flux rate** $\quad$ + **Rate of**

Accumulation $\qquad\qquad$ **IN − OUT** $\qquad\qquad$ **Sources − Sinks**

$$(6.4)$$

or

$$\frac{\partial C}{\partial t} + U\frac{\partial C}{\partial x} = S \qquad\qquad (6.5)$$

EXAMPLE 6.3: *Plug flow reactor with a first-order sink*

Consider a plug flow reactor at steady state, with a first-order sink. What would be the concentration profile versus distance down the reactor?

At steady state with a first-order sink, equation (6.4) becomes

$$U\frac{dC}{dx} = -kC \qquad\qquad (E6.3.1)$$

or

$$\frac{dC}{C} = -\frac{k}{U}dx \qquad\qquad (E6.3.2)$$

which, when integrated, will give

$$\ln\left(C/C_0\right) = -kx/U \qquad\qquad (E6.3.3)$$

or

$$\frac{C}{C_0} = e^{-kx/U} \qquad\qquad (E6.3.4)$$

Then, at the end of the reactor, where $x = L$ and $C = C_L$

$$\frac{C_L}{C_0} = e^{-kL/U} = e^{-kt_r} \qquad\qquad (E6.3.5)$$

where $L/U = t_r$, the residence time of the reactor.

Although there is no true plug flow, there are situations where plug flow is an appropriate assumption to make. For example, the gradients for dissolved oxygen

(DO) and biochemical oxygen demand (BOD) are not large in a river system because none of the sources or sinks have a fast reaction time and the source of oxygen – the atmosphere – is continuous. Thus, mixing is not important when considering DO in a river system. Example 6.4 will demonstrate.

EXAMPLE 6.4: *Development of the Streeter–Phelps (1925) equation for DO sag below a point BOD source in a river (plug flow with a first-order and zero-order source/sink terms)*

Consider the wastewater treatment plant on a river, as illustrated in Figure E6.4.1. The river flows with a cross-sectional mean velocity of U, a discharge of Q_r, a DO concentration of C_0, and a BOD concentration of L_r. The wastewater treatment plant adds an additional BOD of L_w discharging at Q_w. What will be the DO concentration in the river reach downstream?

We will make the following assumptions:

1. The problem is steady state or $\partial C/\partial t = 0$.
2. The river can be simulated as a plug flow system.
3. The bacteria population that accounts for BOD is constant, such that we do not need to include it explicitly in the model.
4. BOD decay is first order.

The fourth assumption is probably the least accurate in low DO systems because BOD decay is actually closer to a Monod type of reaction. With these assumptions, we still have two governing equations to solve. The first is for BOD:

$$U\frac{dL}{dx} = -k_1 L \qquad (E6.4.1)$$

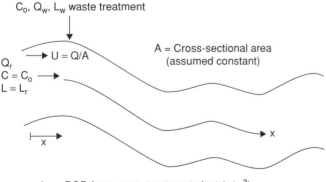

L_w = BOD from waste treatment plant (g/m³)
Q_w = Discharge from waste treatment plant (m³/sec)
L_r = BOD in river
C_0 = DO conc. at x = 0 (cross-sectional mean)
U = Cross-sectional mean velocity

Figure E6.4.1. Illustration of a wastewater treatment plant discharging into a river. BOD, biochemical oxygen demand; DO, dissolved oxygen.

with boundary conditions:

1. $L = L_0 = (Q_r L_r + Q_w L_w)/(Q_r + Q_w)$ at $x = 0$
2. $L = 0$ as $x \Rightarrow \infty$

where k_1 is the rate constant for BOD decay. The second governing equation is for DO:

$$U\frac{dC}{dx} = -k_1 L + k_2(C_E - C) \qquad (E6.4.2)$$

with boundary conditions:

1. $C = C_0$ at $x = 0$
2. $C \Rightarrow C_E$ as $x \Rightarrow \infty$

where k_2 is the reaeration coefficient, k_L/h, k_L is the air–water transfer coefficient for oxygen, h is river depth, and C_E is the DO concentration that is in equilibrium with the atmosphere. Atmospheric transfer of oxygen is included as a source term because DO in the river is averaged over a cross section. Even though oxygen transfer occurs at a boundary, it cannot be considered as anything else but a source term.

The solution to equation (E6.4.1) is similar to that for plug flow with first-order kinetics:

$$L = \beta_0 \, e^{-k_1 x/U} \qquad (E6.4.3)$$

Applying boundary condition 1 to equation (E6.4.3) results in $\beta_0 = L_0$, and

$$L = L_0 \, e^{-k_1 x/U} \qquad (E6.4.4)$$

Now, we need to substitute equation (E6.4.4) into (E6.4.2):

$$\frac{dC}{dx} + \frac{k_2}{U}C = \frac{-k_1 L_0}{U}e^{-k_1 x/U} + \frac{k_2}{U}C_E \qquad (E6.4.5)$$

or, with the operator $\lambda = d/dx$,

$$\lambda + \frac{k_2}{U}C = \frac{-k_1 L_0}{U}e^{-k_1 x/U} + \frac{k_2}{U}C_E \qquad (E6.4.6)$$

Equation (E6.4.6) is a linear ordinary differential equation and has the solution

$$C = C_c + C_p \qquad (E6.4.7)$$

where C_c is the complementary solution, determined from

$$\left(\lambda + \frac{k_2}{U}\right)C_c = 0 \qquad (E6.4.8)$$

or

$$\lambda = \frac{-k_2}{U} \qquad (E6.4.9)$$

Then, C_c has the solution

$$C_c = \beta_1 e^{-k_L x/(hU)} \qquad (E6.4.10)$$

where β_1 is a constant to be determined from boundary conditions. The solution for C_p is chosen based on the right-hand side of equation (E6.4.6) (Kreyszig, 1982):

$$C_p = \beta_2 e^{-k_1 x/U} + \beta_3 \qquad \text{(E6.4.11)}$$

where β_2 and β_3 are determined by substitution into equation (E6.4.6):

$$\frac{-k_1}{U} \beta_2 e^{-k_1 x/U} + \frac{k_2}{U} \beta_2 e^{-k_1 x/U} + \frac{k_2}{U} \beta_3 = \frac{-k_1 L_0}{U} e^{-k_1 x/U} + \frac{k_2 C_E}{U} \qquad \text{(E6.4.12)}$$

Taking the terms with an exponential term to solve for β_2 and the terms without an exponential term to solve for β_3, results in

$$\beta_3 = C_E \qquad \text{(E6.4.13)}$$

and

$$\beta_2 = \frac{-L_0}{\dfrac{k_L}{h\,k_1} - 1} \qquad \text{(E6.4.14)}$$

Then, equation (E6.4.7) becomes

$$C = \beta_1 e^{-k_2 x/U} + C_E - \frac{L_0}{\dfrac{k_2}{k_1} - 1} e^{-k_1 x/U} \qquad \text{(E6.4.15)}$$

Applying the boundary condition that $C = C_0$ at $x = 0$ results in

$$\beta_1 = C_0 - C_E + \frac{L_0}{\dfrac{k_2}{k_1} - 1} \qquad \text{(E6.4.16)}$$

and equation (E6.4.15) becomes

$$C = C_0 e^{-k_2 x/U} + \frac{L_0}{\dfrac{k_2}{k_1} - 1} \left(e^{-k_2 x/U} - e^{-k_1 x/U}\right) + C_E \left(1 - e^{-k_2 x/U}\right) \qquad \text{(E6.4.17)}$$

The solution to equation (E6.4.17) is plotted along with equation (E6.4.4) in Figure E6.4.2, for given values of the independent parameters. When the BOD is high, the BOD decay requires more oxygen than can be supplied through air–water oxygen transfer, so the DO concentration decays over distance. Eventually, the BOD concentration will be low enough and oxygen transfer will be high enough that DO concentration increases. The result is the DO sag curve, first derived by Streeter and Phelps (1925).

C. Complete Mix Reactors in Series

An estuary or river will not be completely mixed, but will have some mixing that goes with the convective transport velocity. Complete mix reactors-in-series respond to a first-order reaction in the manner given in Example 6.5.

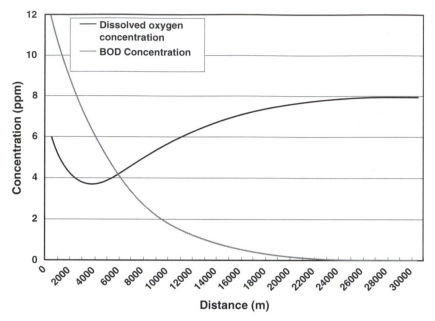

Figure E6.4.2. Dissolved oxygen sag curve for $C_0 = 6$ ppm, $L_0 = 12$ ppm, $C_E = 8$ ppm, $k_L = 10^{-4}$ m/s, $k_1 = 10^{-4}$ m/s, $h = 2$ m, and $U = 0.5$ m/s. BOD, biochemical oxygen demand.

EXAMPLE 6.5: *Steady-state reactors-in-series with a first-order sink*

There are a series of n reactors-in-series, as indicated in Figure E6.5.1. What will be the concentration coming out of the nth reactor if the process is at steady state with a first-order sink?

The mass transport equation for tank 1 is written as

$$C_0 Q - C_1 Q - \forall k C_1 = 0 \qquad (E6.5.1)$$

and for tank 2:

$$C_1 Q - C_2 Q - \forall k C_2 = 0 \qquad (E6.5.2)$$

and for tank n:

$$C_{n-1} Q - C_n Q - \forall k C_n = 0 \qquad (E6.5.3)$$

Now, equation (E6.5.1) has a solution

$$\frac{C_1}{C_0} = \frac{1}{1 + kt_r} \qquad (E6.5.4)$$

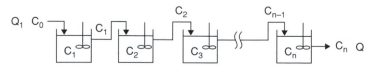

Figure E6.5.1. Illustration of reactors in series.

where $t_r = V/Q$. Equation (E6.5.2) has a solution

$$\frac{C_2}{C_1} = \frac{1}{1 + kt_r} \qquad (E6.5.5)$$

or

$$\frac{C_2}{C_0} = \frac{1}{(1 + kt_r)^2} \qquad (E6.5.6)$$

From equation (E6.5.6) we can see that, for any numbered reactor n, the concentration leaving the reactor, C_n would be

$$\frac{C_n}{C_0} = \frac{1}{(1 + kt_r)^n} \qquad (E6.5.7)$$

In the next example, we will compare a complete mixed reactor with a first-order degradation reaction at steady state for a plug flow reactor and a series of smaller complete mixed reactors.

EXAMPLE 6.6: *Degradation in various systems modeled by one complete mixed reactor, one plug flow reactor, and several complete mixed reactors-in-series*

The compound formaldehyde is biodegrading in several systems simultaneously: (1) a lake, modeled as a complete mix reactor; (2) an estuary, modeled as three complete mixed reactors in series; (3) a large river, modeled as 10 complete mixed reactors in series, and a small stream, modeled as a plug flow reactor. What is the nondimensional reaction/residence time, $k\,t_r$, that is required for each of these systems to reach a degradation of 50%, 90%, 99%, and 99.9%?

The answers to Example 6.6 may be found using equation (E6.3.5), (E6.5.2), and (E6.5.7), and are given in Table E6.6.1.

Table E6.6.1 tells us two things that are important to understanding these reactors. First, it takes significantly more residence time to achieve a given level of degradation in a complete mixed reactor than in a plug flow reactor. This is because the mixed reactor is continually mixing in compound that is not degraded, similar to a lake or reservoir. The plug flow reactor has no mixing of fresh compound

Table E6.6.1: *Nondimensional degradation times, $k\,t_r$, required for various levels of degradation in four systems*

System	Model	Equation for C/C_0	$\Sigma(kt_r)$ to achieve degradataion of:			
			50%	90%	99%	99.9%
Lake	CMR	$= 1/(1 + kt_r)$	1.0	9.0	99.0	999
Estuary	3 CMRs	$= 1/(1 + \Sigma kt_r/3)^3$	0.77	3.5	10.9	27
River	10 CMRs	$= 1/(1 + \Sigma kt_r/10)^{10}$	0.72	2.6	5.9	10
Stream	Plug flow	$= \exp(-kt_r)$	0.61	2.3	4.6	6.9

CMR, complete mixed reactor.
Note: 1 − (degradation %/100) = C/C_0.

with reacted compound, so the entire outflow from a plug flow reactor will have the residence time of the reactor to degrade. Second, the difference between the dimensionless degradation times of the river (modeled with 10 complete mixed reactors) and the smaller stream (modeled as plug flow) is relatively small. There is not much difference between 10 complete mixed reactors and a plug flow reactor for these first-order reactions at steady state. This last observation is true unless the inflow concentration is highly unsteady (i.e., a pulse or a front). It provides the rationale for discretizing streams and rivers into a series of well-mixed cells in the computational transport routines discussed in Chapter 7.

D. Tracer Studies to Determine Reactor Parameters

When we talk about *tracers*, we generally mean conservative tracers with no sources or sinks. This is opposed to *gas tracers*, with gas transfer to the atmosphere, and *reactive tracers*, with a reaction occurring. Tracer studies typically use a conservative tracer, input to the system in a highly unsteady manner, such as a pulse or a front. The pulse and front are typically a more stringent test of the model than a steady-state process with any variety of reactions. Thus, a model that properly simulates the output concentration curve of a pulse or front is assumed to be sufficient for most real conditions with reactions.

When we use a tracer study to develop reactor parameters for an environmental system, we are inherently assuming that the details of the transport processes are not essential to us. All that we have is an input and an output, and any sets of reactors that will simulate the output for a given input are acceptable. What you can learn about the system from a reactor model depends on your understanding of the transport processes and how they are simulated by reactor models.

The reactor combinations that are given here are those that have been found to best simulate tracer studies for environmental systems. A more complete reactor analysis is provided in reactor texts, such as Levenspiel (1962).

1. Complete Mix Reactor

The response of a single complete mix reactor to a pulse or front input is essential to our analysis. A "pulse" would be a Dirac delta in concentration, whereas a front would be a sharp step at a given time to a different concentration. Let us first develop the outflow equation for a front input, illustrated in Figure 6.3, with boundary conditions:

1. At $t < 0$, $C_i = C_0$, and $C = C_0$
2. At $t > 0$, $C_i = 0$

where C_i is the concentration of the inflow. Then, equation (6.3) becomes

$$\forall \frac{dC}{dt} = QC_i - QC \qquad (6.6)$$

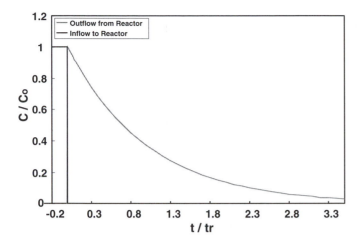

Figure 6.3. Response of a single complete mix reactor to a front in concentration.

Now, separating variables, integrating, and applying boundary condition 1 results in

$$\frac{C}{C_0} = e^{-t/t_r} \tag{6.7}$$

which is also given in Figure 6.3.

The response of a single complete mix reactor to a conservative pulse of mass M at $t = 0$ would have boundary conditions:

1. At $t = 0$, $C = M/\mathcal{V} = C_0$
2. At $t < 0$, $C_i = 0$
3. At $t > 0$, $C_i = 0$

The concentration boundary conditions for C in the reactor are the same as for the front. Then, equation (6.3) again results in

$$\frac{C}{C_0} = e^{-t/t_r} \tag{6.7}$$

that is illustrated in Figure 6.4.

2. Complete Mix Reactors in Series

For a single complete mix reactor, we found that there is little difference in the nondimensional response to a front versus a pulse. This will not be the case for a series of complete mix reactors. Let us assume that we have four complete mix reactors in series, each with a volume $\mathcal{V}_i$ and a residence time, t_{ri}, as illustrated in Figure 6.5. Then, for a pulse input, we have the boundary conditions:

1. At $t = 0$, $C_1 = M/\mathcal{V}_i = C_0$
2. At $t = 0$, $C_2 = 0$
3. At $t = 0$, $C_3 = 0$
4. At $t = 0$, $C_4 = 0$

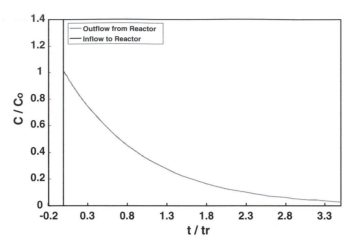

Figure 6.4. Response of a single complete mix reactor to a pulse in concentration.

Solving the four equations for C_1, C_2, and C_3 gives

$$\frac{C_1}{C_0} = e^{-t/t_{ri}} \tag{6.8}$$

$$\frac{C_2}{C_0} = \frac{t}{t_{ri}} e^{-t/t_{ri}} \tag{6.9}$$

$$\frac{C_3}{C_0} = \frac{1}{2} \left(\frac{t}{t_{ri}} \right)^2 e^{-t/t_{ri}} \tag{6.10}$$

and

$$\frac{C_4}{C_0} = \frac{1}{2(3)} \left(\frac{t}{t_{ri}} \right)^3 e^{-t/t_{ri}} \tag{6.11}$$

where $t_{ri} = \mathcal{V}_i/Q$ is the residence time of each model reactor. The solution is given in Figure 6.6.

From equation (6.11), we can see that for n reactors of residence time t_{ri}, the solution to equation (6.3) at the last reactor would be

$$\frac{C_n}{C_0} = \frac{(t/t_{ri})^{n-1} e^{-t/t_{ri}}}{(n-1)!} \tag{6.12}$$

The response given in equation (6.12) is precisely the output that one would get from the standard computational mass transport program that uses control volumes

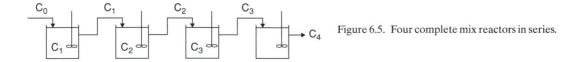

Figure 6.5. Four complete mix reactors in series.

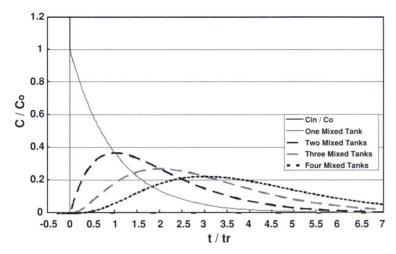

Figure 6.6. Solution to a pulse for four complete mix reactors-in-series.

to a pulse input. This is because each control volume "cell" is assumed to have one concentration at a given moment in time. In other words, each cell is assumed to be a complete mix reactor! This will provide an understanding of the term "numerical diffusion" that is often used when discussing computational transport or computational flow routines. Numerical diffusion primarily results from the fact that the computational models have discretized the solution domain into a series of complete mix reactors.

It is interesting to note that Figure 6.6 approaches a Gaussian distribution as more tanks are placed in the series. In fact, equation (6.12) has a dimensionless variance, which will prove useful in comparisons of tanks-in-series and plug flow with dispersion (which will be discussed later in this chapter).

$$\sigma^2 = \frac{\sigma_t^2}{t_r^2} = \frac{1}{n} \tag{6.13}$$

3. Plug Flow

In Example 6.6, we saw that there is little difference between a plug flow reactor and 10 or more reactors-in-series when the constituent is undergoing a first-order reaction under steady-state conditions. This is *not* true for all circumstances. One example would be a conservative tracer under unsteady boundary conditions, as discussed in this section.

For a conservative tracer with an unsteady input, a plug flow reactor simply acts as a time lag, where the input comes out of the reactor later, precisely as it went in. The examples of a pulse input and a front input at time $t = 0$ are given in Figure 6.7. The only difference between the inflow and the outflow are the times at which the

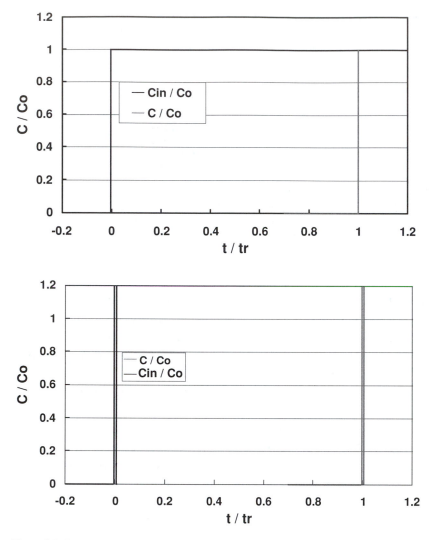

Figure 6.7. Response of a plug flow reactor to a pulse and to a front in concentration at $t = 0$. (*Top*) Front. (*Bottom*) Pulse.

change in concentration occurs. Thus, a plug flow reactor serves similar to a *bypass*, except with a given residence time.

Of course, zero mixing does not exist, and the plug flow reactor is used when the mixing is not important to the process of interest. Mixing can be simulated, however, with a set of complete mix reactors-in-series with a plug flow reactor. If, for example, there was a river or estuary that split into multiple channels, one could model the main channel with a series of complete mix reactors. Then, the parallel channels could be modeled with fewer reactors-in-series and a plug flow reactor to simulate the lower degree of mixing in the smaller side channels. An example of this type of reactor model will be given in Example 6.7.

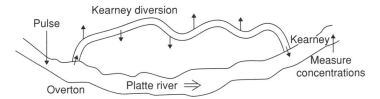

Figure E6.7.1. Illustration of the Platte River with a side channel.

EXAMPLE 6.7: *River bypass with mixing and lag*

A computational model will be developed for numerous water quality parameters in the Platte River, Nebraska. In many locations, this river splits into multiple channels that are joined back together downstream. One significant split is the Kearney Canal diversion, illustrated in Figure E6.7.1, where 20% of the flow splits off into a second river at the city of Overton, only to return 20 km downstream at the city of Kearney. A tracer pulse was put into the river at location $x = 0$ and time $t = 0$, upstream of the diversion. Downstream of the diversion's return, the pulse at location $x = 25$ km is given in Figure E6.7.2. Develop a model for this reach that contains equal size tanks-in-series for the main channel and a similar number of tanks-in-series with the addition of a possible plug flow for the side channel, as illustrated in Figure E6.7.3.

In Figure E6.7.2, the first mass, M_1, corresponds to the pulse that traveled down the main reach, and the second mass, M_2, corresponds to the tracer that traveled along

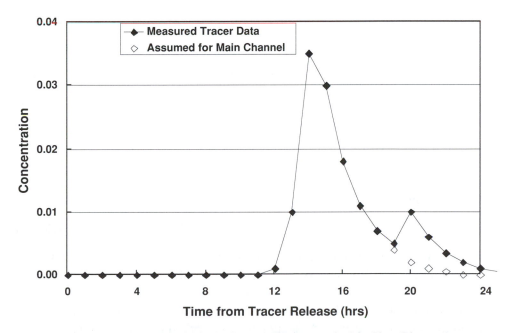

Figure E6.7.2. Results of a tracer pulse in the Overton–Kearney reach of the Platte River.

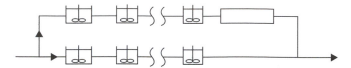

Figure E6.7.3. Tanks-in-series with possible plug flow model of the Platte River.

the diversion. The discharge in the river at a location downstream of the study reach was 100 m³/s.

The information provided in Figure E6.7.2 was computed with the tracer pulse curves applied to the following equations:

$$t_{r_1} = \frac{\sum C_1 t \, \Delta t}{\sum C_1 \, \Delta t} = 15.3 \, \text{hrs} \tag{E6.7.1}$$

$$t_{r_2} = \frac{\sum C_2 t \, \Delta t}{\sum C_2 \, \Delta t} = 21.2 \, \text{hrs} \tag{E6.7.2}$$

$$\sigma_{t1}^2 = \frac{\sum C_1 (t - t_{r_1})^2 \, \Delta t}{\sum C_1 \, \Delta t} = 3.11 \, \text{hrs}^2 \tag{E6.7.3}$$

$$\sigma_{t2}^2 = \frac{\sum C_2 (t - t_{r_2})^2 \, \Delta t}{\sum C_2 \, \Delta t} = 2.36 \, \text{hrs}^2 \tag{E6.7.4}$$

The two pulses are presumed separated by the dashed line given in Figure E6.7.2. For a large number of tanks-in-series (10 or more) the pulse curve is close to a Gaussian probability curve. From equation (6.12), there are

$$n_1 = t_{r1}^2 / \sigma_{t1}^2 = 75 \tag{E6.7.5}$$

tanks-in-series for the main channel of volume

$$V_1 = \frac{Q_1 t_{r1}}{n_1} = 58,750 \, \text{m}^3 \tag{E6.7.6}$$

If the diversion is to have the same number of tanks-in-series, then equation (6.12) gives the mixed tank residence time for the diversion:

$$t_{rm} = \sqrt{n_1} \sigma_{tz} = 13.3 \, \text{hrs} \tag{E6.7.7}$$

with the remainder of the residence time,

$$t_{rp} = t_{r_2}^2 - t_{rm} = 21.2 - 13.3 = 7.9 \, \text{hrs} \tag{E6.7.8}$$

being taken up by a plug flow reactor. The reactor model is given in Figure E6.7.4, and the fit of the tracer pulse curves is given in Figure E6.7.5.

The skew in the fit of the tracer curve in Example 6.7 occurs because the tails are not modeled well. This is a problem with the reactors-in-series model and most computational models as well. A solution to this curve-fit problem will be discussed in the next section on leaky dead zones. For most applications, in transport modeling,

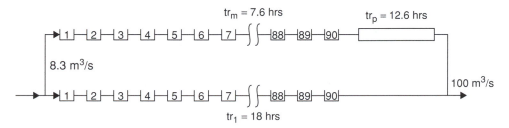

Figure E6.7.4. Reactor model for the Platte River at the bypassed reach.

however, the reactors-in-series models are sufficient. It is primarily with spills, where low concentrations are of interest, that the need for more accurate modeling of the tracer measurement tail is necessary.

4. Leaky Dead Zones

There are many real-life situations resulting in a tracer pulse or front with a long tail, where the pulse or front does not decay nearly as quickly as it rises. The fit of the tanks-in-series to the tracer pulse in Example 6.7 is typical of the "trailing edge" problem. These can be solved by employing a leaky dead-zone model. There are physical arrangements of transport problems where the need of a leaky dead zone seems apparent, such as a side embayment on a lake or river, or a stratified lake where a well-mixed reactor will be used to model the lake. These are illustrated in

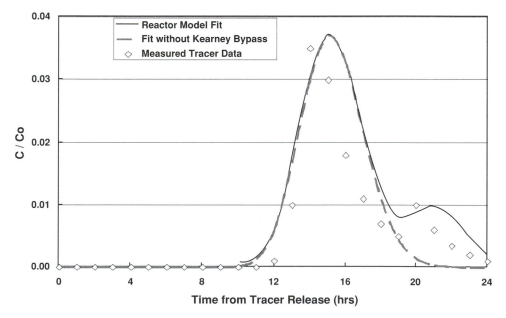

Figure E6.7.5. Tracer curves fit to the field tests at $x = 25$ km, downstream of Kearney, by the reactor model provided in Figure E6.7.4.

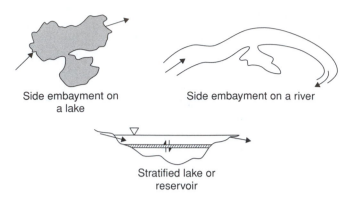

Side embayment on   Side embayment on a river
a lake

Stratified lake or
reservoir

Figure 6.8. Circumstances requiring a leaky dead zone for a model that has minimal bias in fitting a tracer pulse.

Figure 6.8. But, the need of a leaky dead zone is more ubiquitous than these examples imply. To develop a model with a fit, which does not have a strong bias at certain parts of a tracer pulse curve for any reactor or river, for example, a leaky dead zone is a better model.

The concept of a leaky dead zone is illustrated in Figure 6.9. A complete mix reactor is connected to a leaky dead zone through the inflow and outflow discharges to and from the dead zone, Q_d. The dead zone is, in itself, a complete mix reactor, but it is not part of the main flow system with discharge Q. The independent parameters that can be fit to a tracer pulse or front are the volume of the primary complete mix reactor, V_1, the volume of the dead zone, V_d, and the discharge into and out of the dead zone, Q_d.

There are two mass transport equations – the main reactor and the dead-zone reactor – that need to be solved for the reactor combination given in Figure 6.9. These are

$$V_1 \frac{dC}{dt} = C_i Q + C_d Q_d - C(Q + Q_d) + k(C_e - C) \qquad (6.14)$$

and

$$V_d \frac{dC}{dt} = Q_d(C - C_d) + k(C_e - C_d) \qquad (6.15)$$

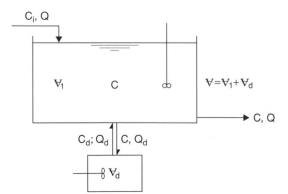

Figure 6.9. Illustration of a complete mix reactor with a leaky dead zone.

where C_d is the concentration in the dead zone, C_i is the inflow concentration, Q_d is the exchange discharge between the main reactor and the dead zone, V_1 is the main reactor volume, and V_d is the dead-zone volume.

We will solve equations (6.14) and (6.15) for a conservative tracer, where $k = o$. Now, if we rearrange equation (6.14) as an explicit equation for C_d, substitute this equation for C_d in equation (6.15) and rearrange this equation, we get

$$\frac{d^2 C}{dt^2} + \left(\frac{1}{t_{d1}} + \frac{1}{t_1} + \frac{1}{t_d} \right) \frac{dC}{dt} + \frac{C}{t_d t_1} = \frac{C_i}{t_d t_1} \tag{6.16}$$

where $t_1 = V_1/Q$, $t_d = V_d/Q_d$, and $t_{d1} = V_1/Q_d$. Now, if we use the $\lambda = d/dt$ operator, equation (6.16) is written as

$$\left[\lambda^2 + \left(\frac{1}{t_{d1}} + \frac{1}{t_1} + \frac{1}{t_d} \right) \lambda + \frac{1}{t_d t_1} \right] C = \frac{C_i}{t_d t_1} \tag{6.17}$$

The solution for equation (6.17) comes from any text on solving ordinary differential equations. It is given as

$$C = \beta_1 e^{\lambda_1 t} + \beta_2 e^{\lambda_2 t} + P(t) \tag{6.18}$$

where the β_1 and β_2 terms are the complimentary solution (if the right-hand side of equation (6.17) is zero), and $P(t)$ is the particular solution, which incorporates the fact that the right-hand side is not truly zero. Now, setting $P(t) = 0$ and solving equation (6.17) for β_1 and β_2 results in

$$\lambda = -\frac{1}{2} \left(\frac{1}{t_{d1}} + \frac{1}{t_1} + \frac{1}{t_d} \right) \left[1 \pm \sqrt{1 - \frac{4}{t_d t_1 \left(\frac{1}{t_{d1}} + \frac{1}{t_1} + \frac{1}{t_d} \right)^2}} \right] \tag{6.19}$$

We will assume that λ_1 and λ_2 indicate the positive and negative signs, respectively, before the square root in equation (6.19). The final solution for $P(t)$ will depend on the inflow concentration, and the determination of β_1 and β_2 will depend on the boundary conditions.

EXAMPLE 6.8: *Reactor model with a complete mix reactor and a leaky dead zone*

A reactor modeled as a complete mix reactor with a leaky dead zone will have a concentration front from $C_i = 0$ to $C_i = C_0$ applied. Develop an equation for the concentration in the outflow of the reactor.

With $C_i = C_0$, the proper choice for $P(t)$ in equation (6.18) is a constant, or $P(t) = \beta_3$. Then, substituting equation (6.18) into (6.17) results in

$$\beta_3 = C_0 \tag{E6.8.1}$$

The boundary conditions for this problem are:

1. At $t = 0$, $C = 0$
2. At $t \Rightarrow \infty$, $C \Rightarrow C_0$
3. At $V_d \Rightarrow 0$, or $t_d \Rightarrow 0$, $C/C_0 = 1 - \exp(-t/t_1)$

Boundary condition 1 results in $\beta_1 + \beta_2 = C_0$.

Boundary condition 2 does not seem to help us much, because both the β_1 term and the β_2 term go to zero as $t \Rightarrow \infty$.

Boundary condition 3, however, when applied results in

$$\lambda_1 \Rightarrow -\infty \tag{E6.8.2}$$

and

$$\lim_{t_d \to 0} \lambda_2 = \frac{-1}{2t_d} \left(1 - \sqrt{1 - 4t_d/t_1}\right) \tag{E6.8.3}$$

If we assign $t_d = \beta t_1$, where β is a small number, equation (E6.8.3) becomes

$$\lim_{t_d \to 0} \lambda_2 = \frac{-1}{2\beta t_1} \left(1 - \sqrt{1 - 4\beta}\right) = -1/t \tag{E6.8.4}$$

if β is sufficiently small. Then,

$$C_0 \left(1 - e^{-t/t_1}\right) = C_0 + \beta_2 e^{-t/t_1} \tag{E6.8.5}$$

Equation (E6.8.5) results in $\beta_2 = C_0$. From boundary condition 1, $\beta_1 = 0$, and the solution to the front is

$$C/C_0 = 1 - e^{\lambda_2 t} \tag{E6.8.6}$$

Equation (E6.8.6) is the solution to a step up in concentration at $t = 0$. With a step down from C_0 to $C = 0$, the solution can be derived from equation (E6.8.6) with a change of variables, or

$$D_{\text{down}} = C_0 - C_{\text{up}} \tag{E6.8.7}$$

Then, the solution to a step down in concentration from $C = C_0$ to $C = 0$ at $t = 0$ is

$$C/C_0 = e^{\lambda_2 t} \tag{E6.8.8}$$

Equation (E6.8.8) is illustrated in Figure E6.8.2. There are two fitted parameters in equation (E6.8.6) or (E6.8.8): t_d and t_1. This is because the third independent parameter is set by the first two parameters:

$$t_{d1} = t_d t_1/(t_r - t_1) \tag{E6.8.9}$$

where

$$t_r = \forall/Q$$

We will now develop the solution to a pulse inflow into a complete mix reactor with a leaky dead zone.

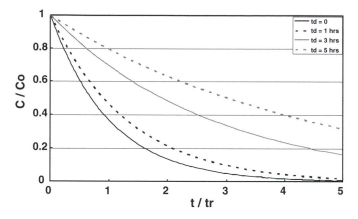

Figure E6.8.2. Solution of a complete mix reactor with a leaky dead zone to a step-down concentration front applied at $t = 0$ with $t_r = 1.2$ hrs and $t_1 = 1$ hr.

EXAMPLE 6.9: *Pulse input into a complete mix reactor with a leaky dead zone*

Equations (6.16) to (6.18) are still applicable to the pulse input. The boundary conditions and the inflow concentration, however, are different. The inflow concentration at $t = 0^+$ will be zero. Thus, in equation (6.17), $\beta_3 = 0$. The new boundary conditions are:

1. At $t = 0^+$, $C = C_0 V/V_1$, where $C_0 = M/V$
2. As $V_d \Rightarrow 0$, $C \Rightarrow C_0 V/V_1 \exp(-t/t_1)$

Boundary condition 1 occurs because the pulse is instantaneously mixed only into the primary reactor in the model. Boundary condition 2 indicates that, as the dead zone disappears, the reactor becomes a standard complete mix reactor. Applying boundary condition 1 gives

$$C_0 V/V_1 = \beta_1 + \beta_2 \tag{E6.9.1}$$

Boundary condition 2 gives

$$\lim_{t_d \to 0} \lambda_2 = -1/t_1 \tag{E6.9.2}$$

and

$$\lim_{t_d \to 0} \lambda_1 = -\infty \tag{E6.9.3}$$

and finally, applying this boundary condition to equation (6.17):

$$\frac{C_0 V}{V_1} e^{-t/t_1} = 0 + \beta_2 e^{-t/t_1} \tag{E6.9.4}$$

or

$$\beta_2 = C_0 V/V_1 \tag{E6.9.5}$$

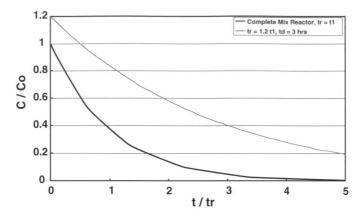

Figure E6.9.1. Solution of a complete mix reactor with and without a leaky dead zone to a pulse input in concentration at $t = 0$.

With this definition for β_2, the initial value of C_0 will generally be greater than 1. Note that equations (E6.9.1) and (E6.9.5) result in $\beta_1 = 0$. The solution to a pulse tracer input to a complete mix reactor with a leaky dead zone is thus

$$\frac{C}{C_0} = \frac{\mathcal{V}}{\mathcal{V}_1}e^{\lambda_2 t} = \frac{t_r}{t_1}e^{\lambda_2 t} \tag{E6.9.6}$$

where λ_2 is given in equation (6.16). Equation (E6.9.6) is illustrated in Figure E6.9.1.

E. Plug Flow with Dispersion

Dispersion is the enhanced mixing of material through spatial variations in velocity. When it is of interest (when we are not keeping track of the three-dimensional mixing), dispersion is typically one or two orders of magnitude greater than turbulent diffusion. The process of dispersion is associated with a spatial mean velocity. The means used in association with diffusion, turbulent diffusion, and dispersion are identified in Table 6.2.

The means by which diffusion (and possibly turbulent diffusion) is combined with a spatial mean velocity to result in dispersion is illustrated in Figure 6.10. A velocity profile over space with mixing due to diffusion (and possibly turbulent diffusion) is

Table 6.2: *Temporal or spatial means and scales used in association with various mixing processes*

Process	Variable representing process	Mean	Scale of mean
Diffusion	Diffusion coefficient	Temporal	Molecular
Turbulent diffusion	Turbulent diffusion coefficient	Temporal	Minutes
Dispersion	Dispersion coefficient	Spatial	Scale of flow

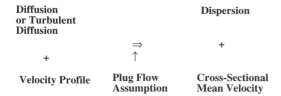

Figure 6.10. Representation of the process by which diffusion or turbulent diffusion is related to dispersion.

combined into a cross-sectional mean velocity and a dispersion coefficient. Without the cross-sectional mean velocity, there *is no* dispersion coefficient.

Dispersion was first developed as a means of dealing with reactors, where there was little interest in the processes creating mixing inside of the reactor, but great need to describe appropriately the output from the reactor. The physics of the mixing process is lost in the conversion to a spatial mean velocity profile, but the end result can still be modeled by dispersion. Dispersion is, therefore, placed in this chapter on reactor mixing assumptions with mixed tank and plug flow reactors, where reality is replaced by ideal systems that can be combined to provide a mixing result similar to the real one.

A similar spatial mean velocity (bulk mean velocity) is used for the plug flow reactor model. Thus, plug flow with dispersion is a natural match, where the mixing that truly occurs in any reactor or environmental flow is modeled as dispersion. This is the model that will be applied to utilize dispersion as a mixing model.

1. Dispersion in Laminar Flow

Any flow with a nonuniform velocity profile will, when spatial mean velocity and concentration are taken, result in dispersion of the chemical. For laminar flow, the well-described velocity profile means that we can describe dispersion analytically for some flows. Beginning with the diffusion equation in cylindrical coordinates (laminar flow typically occurs in small tubes):

$$\frac{\partial C}{\partial t} + U\frac{\partial C}{\partial x} = D\left(\frac{\partial^2 C}{\partial r^2} + \frac{1}{r}\frac{\partial C}{\partial r} + \frac{\partial^2 C}{\partial x^2}\right) + S \tag{6.20}$$

where the x-coordinate is aligned with the flow velocity and $v = w = 0$. We will outline the development of a dispersion coefficient for a fully developed laminar pipe flow. This means that we are far enough from the entrance that the velocity profile is essentially in equilibrium with the loss of pressure along the pipe. This flow has the velocity profile developed in Example 4.3. Combining equations E4.3.5 and E4.3.6 gives

$$u = U_{\max}(1 - r^2/R_t^2) \tag{6.21}$$

where R_t is the tube radius as illustrated in Figure 6.11.

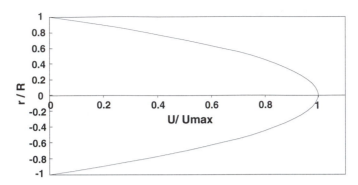

Figure 6.11. The velocity profile in a fully developed tubular flow is a paraboloid.

To convert equation (6.21) to cross-sectional mean values, we will assign

$$\hat{C} = \frac{1}{\pi R^2} \int_0^{R_t} C \, 2\pi r \, dr \tag{6.22}$$

$$U = \frac{1}{\pi R_t^2} \int_0^{R_t} u \, 2\pi r \, dr = \frac{U_{max}}{2} \tag{6.23}$$

Then, in equation (6.20):

$$C = \hat{C} + C'(r) \tag{6.24}$$

and

$$u = U + u'(r) \tag{6.25}$$

Now, if we equate equation (6.21) and (6.25), we get

$$u'(r) = U_m \left(\frac{1}{2} - \frac{r^2}{R_t^2} \right) \tag{6.26}$$

Substituting equations (6.24) to (6.26) into 6.20:

$$\frac{\partial \hat{C}}{\partial t} + \frac{\partial C'}{\partial t} + (U + u') \frac{\partial (\hat{C} + C')}{\partial x} = D \left[\frac{\partial^2 (\hat{C} + C')}{\partial r^2} \right] + \frac{1}{r} \frac{\partial (\hat{C} + C')}{\partial r} + \frac{\partial^2 (\hat{C} + C')}{\partial x^2} + S \tag{6.27}$$

An order of magnitude analysis will tell us that $\partial^2 / \partial r^2 \gg \partial^2 / \partial x^2$. In addition, by definition, $\partial \hat{C} / \partial r = 0$. Now, if we put everything that is known on the left-hand side of equation (6.27), the result will be

$$\frac{\partial \hat{C}}{\partial t} + U \frac{\partial \hat{C}}{\partial x} = -\left(\frac{\partial C'}{\partial t} + U \frac{\partial C'}{\partial x} + u' \frac{\partial \hat{C}}{\partial x} + u' \frac{\partial C'}{\partial x} \right) + D \frac{\partial^2 C'}{\partial r^2} + \frac{D}{r} \frac{\partial C'}{\partial r} + S$$

$$= D_L \frac{\partial^2 \hat{C}}{\partial x^2} + S \tag{6.28}$$

The second equality in equation (6.28) is a definition of longitudinal dispersion coefficient, D_L. Taylor (1953) assumed that some of the terms in equation (6.28) would cancel and that longitudinal convective transport would achieve a balance with transverse diffusive transport. He then solved the second equality in equation (6.28), for a fully developed tubular flow, resulting in the relation

$$D_L = \frac{R_t^2 U^2}{48 D} = \frac{d^2 U^2}{192 D} \tag{6.29}$$

A similar relation can be developed for laminar flow down an inclined plate:

$$D_L = \frac{32 h^2 U^2}{945 D} \tag{6.30}$$

The longitudinal dispersion coefficient is proportional to the square of the flow scale (d or h), proportional to the square of the velocity scale (U), and *inversely* proportional to the diffusion coefficient. The greater the diffusion, the less severe the spread of the chemical by the velocity profile because of local mixing and the smaller the longitudinal dispersion coefficient. This result may seem illogical, but can be explained by the following: Longitudinal dispersion describes mixing only in terms of the cross-sectional mean concentration, and transverse mixing actually slows down longitudinal dispersion. The governing equation (6.28), does not concern itself with cross-sectional mixing issues.

Knowing the relations given in equations (6.29) and (6.30), we no longer need the more cumbersome middle portion of equation (6.27), and we can work to solve the equation

$$\frac{\partial \hat{C}}{\partial t} + U \frac{\partial \hat{C}}{\partial x} = D_L \frac{\partial^2 \hat{C}}{\partial x^2} + S \tag{6.31}$$

In expressing the equations for longitudinal dispersion, we will drop the "hat" above the cross-sectional mean values. It will be assumed for the remainder of Chapter 6, that if D_L is involved, we are discussing cross-sectional mean concentrations.

2. Dispersion in Turbulent Flow

The dispersion that occurs in turbulent flow can also be calculated, as long as the velocity profile is given. This was done by Taylor (1954) for a tubular flow and by Elder (1959) for a two-dimensional, open-channel flow. Both investigators assumed that a logarithmic velocity profile would apply in the entire flow field. The logarithmic velocity profile, as you may remember from Chapter 5, applies in the region where shear stress can be assumed constant. Nevertheless, it is not a bad assumption for a fully developed turbulent flow field, because the locations where the logarithmic profile applies are those with the greatest change in velocity.

For a fully developed tubular flow, assuming a logarithmic velocity profile, Taylor derived the equation

$$D_L = 5.05 \, d u_* \tag{6.32}$$

where u_* is the shear velocity at the wall. Elder (1959) derived the following equation for a two-dimensional, open-channel flow:

$$D_L = 5.93\,hu_*$$ (6.33)

It is interesting to compare equations (6.32) and (6.33) with those for a fully developed laminar flow, equations (6.29) and (6.30). In Example 5.1, we showed that eddy diffusion coefficient in a turbulent boundary layer was linearly dependent on distance from the wall and on the wall shear velocity. If we replace the diffusion coefficient in equation (6.30) with an eddy diffusion coefficient that is proportional to hu_*, we get

$$D_L \frac{h^2 U^2}{hu_*}$$ (6.34)

Noting that for a given boundary roughness we can generally say that $U \sim u_*$, equation (6.34) becomes

$$D_L \sim hu_*$$ (6.35)

which is what we have in equation (6.33).

The relations developed for longitudinal dispersion coefficient are given in Table 6.3. The experimental results in rivers tend to have a large range because of the variety of lateral velocity profiles that exist in natural rivers and streams.

Table 6.3: *Relationships for longitudinal dispersion coefficient in pipes and channels developed from theory and experiments (after Fisher et al., 1979)*

Flow conditions	D_L	Notation
Laminar flow in a pipe (Taylor, 1953)	$\dfrac{R_t^2 U^2}{48D}$	R_t = radius of tube
Laminar flow down an inclined plate	$\dfrac{32h^2 U^2}{945D}$	U = cross-sectional mean velocity
Laminar flow-linear velocity profile (Couette flow)	$\dfrac{V_p^2 \Delta_z^2}{120D}$	D = diffusivity h = depth V_p = velocity of upper plate Δ_z = spacing of plates
Turbulent flow in a pipe, assuming logarithmic velocity profile (Taylor, 1954)	$10.1\,Ru_*$	u_* = shear velocity
Turbulent flow down an inclined plate, assuming logarithmic velocity profile (Elder, 1959)	$5.93\,hu_*$	
Open flume (experimental)	$8\text{–}400\,hu_*$	
Canals (experimental)	$8\text{–}20\,hu_*$	
Rivers (experimental)	$8\text{–}7500\,hu_*$	

F. Solutions to Transport with Convection

Diffusive transport with convection occurring simultaneously can be solved more easily if we orient our coordinate system properly. First, we must orient one axis in the direction of the flow. In this case, we will choose the x-coordinate so that u is nonzero and v and w are zero. Second, we must assume a uniform velocity profile, $u = U = $ constant with y and z. Then, equation (2.33) becomes

$$\frac{\partial C}{\partial t} + \frac{U}{R}\frac{\partial C}{\partial x} = \frac{D}{R}\left(\frac{\partial^2 C}{\partial x^2} + \frac{\partial^2 C}{\partial y^2} + \frac{\partial^2 C}{\partial z^2}\right) + \frac{S}{R} \qquad (6.36)$$

where S is a source or sink term other than adsorption or desorption and R is the retardation coefficient. In equation (6.36), we are assuming that chemical reaction does not take place on the surface of a solid (i.e., while the chemical is sorbed to solids). Now we will convert our Eulerian (fixed) coordinate system to one that moves (Lagrangian) with velocity U/R, and assign an independent variable

$$x^* = x - Ut/R \qquad (6.37)$$

such that $x^* = 0$ at $x = Ut/R$. The response of the system to a conservative pulse, given in Figure 6.12, indicates that in the Lagrangian coordinate system specified, there is no convection term, only diffusion. Then equation (6.36) becomes

$$\frac{\partial C}{\partial t} = \frac{D}{R}\left(\frac{\partial^2 C}{\partial x^{*2}} + \frac{\partial^2 C}{\partial y^2} + \frac{\partial^2 C}{\partial z^2}\right) + \frac{S}{R} \qquad (6.38)$$

similar to equation (E2.4.1) for unsteady diffusion in three dimensions without convection.

1. Determination of Dispersion Coefficient from Tracer Clouds

The one-dimensional mass transport equation for plug flow with dispersion, and a retardation coefficient of 1, is

$$\frac{\partial C}{\partial t} + U\frac{\partial C}{\partial x} = D_L\frac{\partial^2 C}{\partial x^2} + S \qquad (6.39)$$

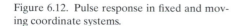
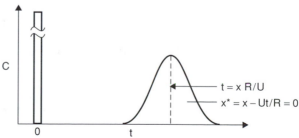

Figure 6.12. Pulse response in fixed and moving coordinate systems.

We will convert our fixed coordinate system to a coordinate system moving at velocity U through the change of variables, $x^* = x - Ut$. Then, equation (6.39) is given as

$$\frac{\partial C}{\partial t} = D_L \frac{\partial^2 C}{\partial x^{*2}} + S \tag{6.40}$$

Now, if we are determining the dispersion coefficient through the use of a pulse tracer cloud, the boundary conditions are those of a Dirac delta:

1. At $t = 0$, a pulse of mass M is released at $x^* = 0$
2. As $t \Rightarrow \infty$, $C \Rightarrow 0$

The similarity between equation (6.40) and (E2.2.2), and the similar boundary conditions means that we can use the solution to equation (E2.2.2) [equation (E2.2.3)] with our new variables:

$$C = \frac{M/A}{\sqrt{4\pi \, D_L t}} \, e^{\frac{-(L-Ut)^2}{4D_L t}} \tag{6.41}$$

or, in dimensionless variables,

$$\frac{C}{C_0} = \frac{C}{M/(AL)} = \frac{1}{\sqrt{4\pi \, Cou/Pe}} \, \exp\left[-\frac{(1 - Cou)^2}{4 \, Cou/Pe}\right] \tag{6.42}$$

where $Cou = Ut/L = t/t_r$ is a Courant number and $Pe = UL/D_L$ is a Peclet number. Comparing equation (6.42) to a Gaussian probability distribution,

$$P(\phi) = \frac{1}{\sigma\sqrt{2\pi}} \, \exp\left(-\frac{\phi^2}{2\sigma^2}\right)$$

we can see that the plug flow with dispersion, when $Pe > 10$, can be fit to a Gaussian distribution in terms of C/C_0 and $(1 - \theta)$, with the relationships

$$\sigma^2 = \frac{\sigma_t^2}{t_r^2} = 2 \, Cou/Pe \tag{6.43}$$

where

$$\sigma_t^2 = \frac{\int_0^\infty t^2 \, C/C_0 \, dt}{\int_0^\infty C/C_0 \, dt} - t_r^2 \tag{6.44}$$

and

$$t_r = \frac{\int_0^\infty t \, C/C_0 \, dt}{\int_0^\infty C/C_0 \, dt} \tag{6.45}$$

In addition, if we could measure the tracer cloud over distance at one time, we would use the relation

$$\sigma^2 = \frac{\sigma_x^2}{U^2 \, t_r^2} \tag{6.46}$$

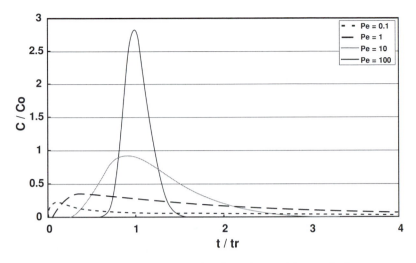

Figure 6.13. Response of the plug flow with dispersion model to a pulse input.

where

$$\sigma_x^2 = \frac{\int_0^\infty x^2 \, C/C_0 \, dx}{\int_0^\infty C/C_0 \, dx} - \bar{x}^2 \quad \text{and} \quad \bar{x} = \frac{\int_0^\infty x \, C/C_0 \, dx}{\int_0^\infty C/C_0 \, dx} \tag{6.47}$$

The response of a plug flow with dispersion model to a pulse input, equation (6.42), is given in Figure 6.13 for various values of the Peclet number, $Pe = UL/D_L$. Equation (6.41) will be applied to the analysis of a pulse with dispersion in a reactor in Example 6.10.

EXAMPLE 6.10: *Air-stripping tower with first-order degradation, modeled as plug flow, plug flow with dispersion, and mixed tanks-in-series reactors*

The volatile organic compound, trichloroethylene, the groundwater pollutant in the novel "Class Action," is present in groundwater pumped into an air stripper. Assume that the volatilization is first-order (does not approach equilibrium with the air in the air stripper), with a first-order rate coefficient ($k_L a$) of 0.307 min^{-1} that has been determined from gas tracer tests on the air-stripping tower. A conservative tracer pulse placed in the inflow of the tower had the outflow concentration given in Table E6.10.1. What is the fraction of trichloroethylene remaining if the air-stripping tower is modeled as a plug flow reactor, a plug flow reactor with dispersion, and complete mix tanks-in-series?

The air-stripping tower, illustrated in Figure E6.10.1, provides the air–water contact area, either through a porous medium that is unsaturated with water, through bubbles rising through the water or both. The polluted water comes in at the top (C_0), and cleaner water comes out of the bottom of the stripping tower, while clean air comes in at the bottom, and air with trichloroethylene in it comes out the top of the tower. This

Table E6.10.1: *Time response of the air-stripping tower output after a conservative tracer pulse input at $t = 0$*

Time (min)	Tracer concentration (g/m^3)
0	0
5	3
10	5
15	5
20	4
25	2
30	1
35	0

is a counterflow configuration, because the water and air are (on the average) flowing in opposite directions. The air treatment out of the top of the stripper depends on the concentrations and air quality regulations in the area.

Plug Flow Reactor. The plug flow reactor model requires only a residence time (t_r). So, the tracer cloud is used to determine t_r:

$$t_r \cong \frac{\sum C_t t \Delta t}{\sum C_t \Delta t} = \frac{5(3) + 10(5) + 15(5) + 20(4) + 25(2) + 30(1)}{3 + 5 + 5 + 4 + 2 + 1} \qquad (E6.10.1)$$

or $t_r = 15$ min.

Now, the equation for first-order degradation in plug flow reactors with a first-order reaction is given in equation (E6.3.5):

$$\frac{C_f}{C_0} = e^{-k_L a \, t_r} \qquad (E6.3.5)$$

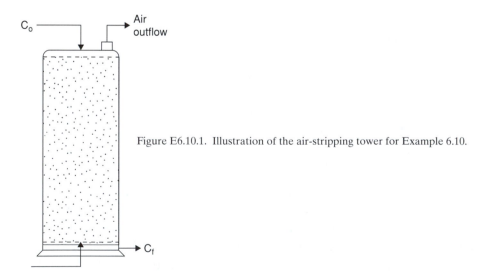

Figure E6.10.1. Illustration of the air-stripping tower for Example 6.10.

where k becomes k_L for our air-stripping tower. Then,

$$\frac{C_f}{C_0} = e^{-0.307(15)} = 0.010 \tag{E6.10.2}$$

The plug flow reactor model predicts that there will be 1% of the trichloroethylene remaining at the bottom of the stripping tower.

Plug Flow Reactor with Dispersion. The residence time is still 15 min. The plug flow with a dispersion model gives equation (6.43):

$$\sigma^2 = \frac{\sigma_t^2}{t_r^2} = 2\,Cou/Pe \tag{6.43}$$

Applying equation (6.44) results in

$$\sigma_t^2 \cong \frac{\sum t^2 \, C_t \, \Delta t}{\sum C_t \, \Delta t} - t_r^2 = \frac{5{,}450}{20} - 15^2 \tag{E6.10.3}$$

and

$$\sigma^2 = \frac{5{,}450}{20(15)^2} - 1 = 0.211 \tag{E6.10.4}$$

However, the dependence of D_L on the variance is complicated by $Cou = t/t_r$, being a function of time. For simplicity, we will assume that Cou in equation (E6.10.4) is equal to 1.0. Then,

$$Pe = 9.4 \tag{E6.10.5}$$

The assumption that $Cou = 1$ in equation (6.43) is really only accurate when $Pe > 10$. The only way to apply this tracer curve to the plug flow with dispersion model while $Cou \neq 1$ would be to route each portion of the tracer curve through the reactor. With $Pe = 9.4$, this solution will be close, although still an approximation.

Now, we need a solution to the plug flow with dispersion model for steady-state operation of an air-stripping tower. The mass transport equation for this situation, assuming minimal trichloroethylene builds up in the bubble, is

$$U\frac{dC}{dx} = D_L\frac{d^2C}{dx^2} - k_L a C \tag{E6.10.6}$$

We can make this equation dimensionless, to match our dimensionless variables, by assigning

$$\hat{C} = C/C_0 \quad \text{or} \quad C = C_0/\hat{C}$$
$$X = x/L \quad \text{or} \quad x = xL, \text{ and}$$
$$St = k_L a\,L/U, \text{ a Stanton number.}$$

Then, equation (E6.10.6) becomes

$$\frac{1}{Pe}\frac{d^2\tilde{C}}{dX^2} - \frac{d\tilde{C}}{dX} - St\,\tilde{C} = 0 \tag{E6.10.7}$$

with boundary conditions:

1. At $X = 0$, $C = 1$
2. As $St \Rightarrow \infty$, $C \Rightarrow 0$

Using the operator, $\lambda = d/dX$, equation (E6.10.7) becomes

$$\left(\frac{1}{Pe}\lambda^2 - \lambda - St \right) \tilde{C} = 0 \qquad \text{(E6.10.8)}$$

Once again, we can use the standard solution technique for ordinary differential equations (Kreyszig, 1982), resulting in a solution,

$$\tilde{C} = \beta_1 e^{\lambda_1 X} + \beta_2 e^{\lambda_2 X} \qquad \text{(E6.10.9)}$$

where the quadratic formula can be applied to equation (E6.10.8):

$$\lambda = \frac{-b}{2a}\left(1 \pm \sqrt{1 - 4ac/b} \right) \qquad \text{(E6.10.10)}$$

with

$$a = 1/Pe \qquad b = -1 \qquad c = St$$

We will assign λ_1 to be the positive sign and λ_2 to be the negative sign in equation (E6.10.10). To determine β_1 and β_2, we will apply our boundary conditions: boundary condition 2 results in $\beta_1 = 0$. Boundary condition 1 results in $\beta_2 = 1$. Then, equation (E6.10.9) becomes

$$\tilde{C}_{|x=L} = \frac{C_f}{C_0} = \exp\left[-\frac{Pe}{2}\left(1 - \sqrt{1 + 4\,St/Pe} \right) \right] \qquad \text{(E6.10.11)}$$

With $Pe = 9.4$ and $St = 0.307(15) = 4.60$, equation (E6.10.11) gives

$$\frac{C_f}{C_0} = 0.034 \qquad \text{(E6.10.12)}$$

The plug flow with dispersion model results in a degradation to 3.4% of the inflow trichloroethylene concentration. This is significantly different than the plug flow model (1.0%). It is also a more accurate solution. Whether it is the tail of a tracer pulse or a reaction that approaches complete degradation, one needs to be careful about applying the plug flow model when low concentrations, relative to the inflow, are important.

Complete Mix Reactors-in-Series. The number of complete mix reactors-in-series that should be applied to fit the tracer curve can be found from an application of equation (6.13):

$$n = \frac{1}{\sigma^2} = \frac{1}{0.211} = 4.76 \qquad \text{(E6.10.13)}$$

There is no reason to round up the number of reactors-in-series, because we can take an exponent to the 4.76 power. Now, if we apply our 4.76 reactors to the steady-state operation of a stripping tower with first-order degradation, equation (E6.5.7) gives

$$\frac{C_f}{C_0} = \frac{1}{(1 + St/n)^n} = 0.040 \tag{E6.5.7}$$

The complete mix reactor-in-series model resulted in degradation to 4% of the initial trichloroethylene concentration.

Degradation to 4% of the inflow concentration compares with the 3.4% for plug flow with dispersion and 1% for the plug flow model. Routing the tracer curve through the reactor though assigning a residence time to each Δt increment is probably the most accurate and most time-consuming technique. This routing process resulted in a degradation to 4.7% of the inflow concentration. Thus, of the three models, the complete mix reactors-in-series performed the best in this case.

Both the tanks-in-series and the plug flow with dispersion models have an effective dispersion that we have related to the variance of the tracer cloud:

$$\sigma^2 = \frac{1}{n} = \frac{2}{Pe} \tag{6.48}$$

The models' response to a pulse input, however, is not close to equivalent until we get to approximately five tanks-in-series. Recall that, with one complete mix tank, we cannot distinguish between a pulse and a front. The output looks the same. Levenspiel (1962) has developed a relation to include the "end effects" of reactors (i.e., there is no dispersion across the end of the reactor):

$$\frac{1}{n} = \frac{2}{Pe} - \frac{2}{Pe^2}(1 - e^{-Pe}) \tag{6.49}$$

This is shown in Figure 6.14, where the tanks-in-series model with and without end effects is compared with plug flow with dispersion. Above a Peclet number of 10,

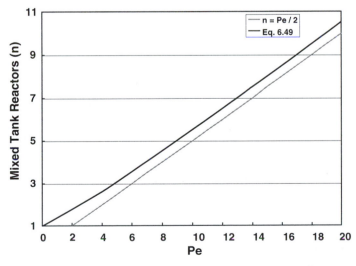

Figure 6.14. Comparison of the longitudinal dispersion at equal σ^2 from the tanks-in-series and plug flow with dispersion models.

there is little difference between the response of the mixed-tanks-in-series and plug flow with dispersion models.

We will now use an example column experiment to determine the retardation coefficient for a given soil.

EXAMPLE 6.11: *Determination of retardation coefficient*

As part of a forensic investigation of a continuous Malathion spill, you need to determine the retardation coefficient of the soil at the site for Malathion. You have decided to do so in a column experiment with the soil (illustrated in Figure E6.11.1). Also given in the figure are the results of a pulse test with the nonsorptive tracer, chloride, and the results of a pulse test with Malathion. What is the retardation coefficient, R?

Chloride:

$$t_r = t_{r1} = 20 \, \text{min}$$
$$\sigma_{t1} = 3 \, \text{min}$$

Malathion:

$$t_r = t_{r2} = 400 \, \text{min}$$
$$\sigma_{t2} = 66 \, \text{min}$$

This test can result in a retardation coefficient from a comparison of both residence times and the variance.

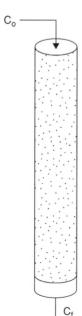

Figure E6.11.1. Illustration of the column test for retardation coefficient and results of the tracer tests.

Residence Times. The residence time of the chloride can be given as $t_{r1} = L/U$, and the residence time of the Malathion can be given as $t_{r2} = RL/U$. Thus,

$$R = \frac{t_{r2}}{t_{r1}} = \frac{400 \text{ min}}{20 \text{ min}} = 20 \qquad \text{(E6.11.1)}$$

Variance of Tracer Curves. Equation (6.43) provides the relationships

$$\sigma^2 = \sigma_t^2/t_r^2 = 2D_L \, Cou/(UL).$$

Therefore, assuming that $Cou = 1$,

$$\sigma_{t2}^2 = \frac{2D_L}{LR}\frac{R}{U}t_{r2}^2 = \frac{2D_L}{UL}t_{r2}^2 = \sigma_{t1}^2\frac{t_{r2}^2}{t_{r1}^2} = R^2\,\sigma_{t1}^2 \qquad \text{(E6.11.2)}$$

or

$$R = \sigma_{t2}/\sigma_{t1} = 66 \text{ min}/3 \text{ min} = 22 \qquad \text{(E6.11.3)}$$

Our two means of determining the retardation coefficient in the column gave $R = 20$ and $R = 22$. We can also check whether the organic carbon content of the soil fits what is generally known from the literature. First, there is the relation for R:

$$R = 1 + \frac{\rho_B}{\varepsilon}K_d \qquad \text{(E6.11.4)}$$

Second, we have the equation from Karikhoff et al. (1979):

$$k_d = 0.41 \left(\frac{\text{cm}^3}{g}\right) f\, K_{ow} \qquad \text{(E6.11.5)}$$

Combining equations (E6.11.4) and (E6.11.5) results in an equation for the organic fraction:

$$f = \frac{(R-1)\varepsilon}{0.41\,\rho_B(g/\text{cm}^3)K_{ow}} = 0.030 - 0.033 \qquad \text{(E6.11.6)}$$

For Malathion, $K_{ow} = 230$. From soil tests, $\varepsilon = 0.3$ and $\rho_B = 2.0 g/\text{cm}^3$. Then, equation (E6.11.6) gives $f = 0.030 - 0.033$. This number is about right for the organic-rich soil that was tested. Thus, we know that our column tests are of the right order. One further test would be to determine the total organic carbon of the soil and compare that value with f.

2. Dispersion in Groundwater Flow

Dispersion in a flow through a porous media occurs due to heterogeneity in the media (i.e., the conductivity of the soil varies with space). This is shown on three levels in Figure 6.15. On the particle scale, a thread of tracer will be split a number of times as it moves through the media. Each split of the tracer thread will move through the media at a speed corresponding to the resistance that it encounters. If you take a number of tracer threads coming out of the media at different times and collected them in an outlet pipe, what you would see at the end of the pipe would be a dispersed

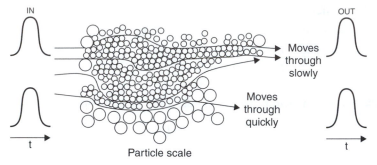

IN
OUT

Moves
through
slowly

Moves
through
quickly

t

t

Particle scale

Two flow paths on the particle scale move through the media at different rates.

On a larger level, fingering is caused by layered beds with a low k
(conductivity) and lenses with a high k

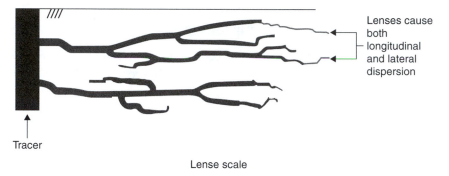

Lenses cause
both
longitudinal
and lateral
dispersion

Tracer

Lense scale

On a still-larger spatial and temporal scale, a "dispersion" cloud

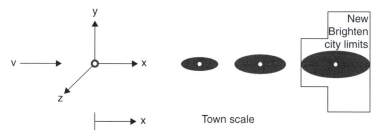

y

v

x

z

x

New
Brighten
city limits

Town scale

Figure 6.15. Illustration of dispersion in groundwater flow at various scales.

pulse. This dispersion would be much greater than the diffusion that would occur.
A lateral dispersion would also occur because the media would move some of the
tracer threads laterally. Thus, in groundwater flow, there generally is longitudinal and
lateral dispersion created by the character of the media.

On a larger scale, a similar process can occur. Fingering of the tracer is created by
layered beds with a low conductivity and lenses with a high conductivity. The tracer
that ends up in a lense travels at a relatively high speed. Those tracer molecules
will reach the measuring point much sooner than the tracer molecules stuck in the
low conductivity beds, thus creating longitudinal dispersion. As the lenses are not all
parallel to each other, they will also create a lateral dispersion of the tracer.

Finally, on a still larger scale, about the scale of a small town, there will be isopleths (lines of constant concentration) of our tracer that look something like those given in the last illustration of Figure 6.15. If enough particle and lense effects have occurred with an apparent randomness to our tracer cloud, then the cloud disperses in a manner similar to that illustrated.

The mass transport equation (2.3.7) is then written as

$$\frac{\partial C}{\partial t} = \frac{(D + D_x)}{R} \frac{\partial^2 C}{\partial x^2} + \frac{(D + D_y)}{R} \frac{\partial^2 C}{\partial y^2} + \frac{(D + D_z)}{R} \frac{\partial^2 C}{\partial z^2} \tag{6.50}$$

where D_x, D_y, and D_z are the dispersion coefficients in the x-, y-, and z-directions, respectively.

Let us assume that we are developing an approximate solution to Example 2.3, except that the barrel had been placed near the shore, where there is still a significant horizontal pressure gradient causing groundwater flow. In that case, we can determine that the appropriate solution would be

$$C = \frac{MR^{1.5}}{8(\pi^3 t^3 D_x D_y D_z)^{0.5}} \exp\left(\frac{-Rx^{*2}}{4D_x t} \frac{-Ry^2}{4D_y t} \frac{-Rz^2}{4D_z t}\right) \tag{6.51}$$

The dispersion coefficients are dependent on the character of the media and the flow velocity. It is difficult to predict these coefficients within an order of magnitude, so they are normally measured or fit to measured data in the field. For example, if a concentration variance in the x-, y-, and z-directions can be measured in response to a pulse release, then equation (6.51) could be applied to obtain

$$D_x = 9\sigma_x^2/(2t) \tag{6.52a}$$

$$D_y = 9\sigma_y^2/(2t) \tag{6.52b}$$

$$D_z = 9\sigma_z^2/(2t) \tag{6.52c}$$

where $9\sigma_x^2$, σ_y^2, and $9\sigma_z^2$ are the variance of concentration in the x-, y-, and z-directions, respectively. They are given by the equations

$$\sigma_x^2 = \frac{\int x^2 C \, dx}{\int C \, dx} - \left(\frac{\int x C \, dx}{\int C \, dx}\right)^2 \tag{6.53a}$$

$$\sigma_y^2 = \frac{\int y^2 C \, dx}{\int C \, dx} \tag{6.53b}$$

$$\sigma_x^2 = \frac{\int z^2 C \, dx}{\int C \, dx} \tag{6.53c}$$

The last term in equation (6.53a) is the distance to the center of mass. In the diffusion equation, it is equal to Ut.

Dispersion Coefficients in Groundwater Flow.

In a uniform media of particles, the longitudinal dispersion coefficient, D_L, and the transverse dispersion coefficient, D_t, are both functions of the grain diameter and velocity. (In our previous example, D_L

was D_x, and D_t would indicate D_y and D_z.) The relevance of longitudinal and transverse dispersion relative to diffusion may therefore be *very roughly* characterized by a Peclet number, *Pe*:

$$Pe = Ud/D \qquad (6.54)$$

where d is the grain diameter and U is a bulk velocity, Q/A, where the cross section includes porous media. If the consideration is not of uniform media, but is for a heterogeneous region of high and low groundwater flow permeability, then the appropriate length scale would be the size of these permeability regions normal to the flow. The characterization is (Freeze and Cherry, 1979):

$$D_L/D \sim Pe \qquad (6.55)$$

Koch and Brady (1985) have characterized transverse dispersion coefficient as a fraction of longitudinal dispersion coefficient:

$$Dt/DL \sim 0.1 \qquad (6.56)$$

Thus, longitudinal dispersion coefficient is roughly 10 times the value of transverse dispersion coefficient in a uniform media.

In the field, however, all media are heterogeneous, resulting in far greater dispersion than in uniform porous media. Because of the heterogeneities, the velocity profile can be highly variable over long distances, creating a much greater dispersion. Gelhar et al. (1992) provided a plot of field data that can be manipulated to result in an equation that applies between a scale of 1 and 100,000 m:

$$Pe_{D_L} = \frac{UL}{D_L} = 2.5 \times 10^{1\pm 1.4} \qquad (6.57)$$

where L is the horizontal scale of the measurement. The field data are given in Figure 6.16, with the curves of equation (6.57).

EXAMPLE 6.12: *Atrazine spill into an irrigation well (three-dimensional dispersion with convection)*

Three kilograms of 1,000 g/m^3 atrazine pesticide is accidentally dumped down an old farm irrigation well, placed to pump water out of porous sandstone. Estimate the movement of the atrazine plume over time and the concentrations in the plume. (See Figure E6.12.1.)

Given:

$$U = 10^{-4} \text{m/s} \qquad d(\text{sandstone}) \cong 1 \text{ mm}$$
$$f = 10^{-4} (\text{not soil}) \qquad D = 10^{-10} \text{m}^2/\text{s}$$
$$\varepsilon = 0.3 \qquad \rho_b/\varepsilon = 6 \text{ g/cm}^3$$

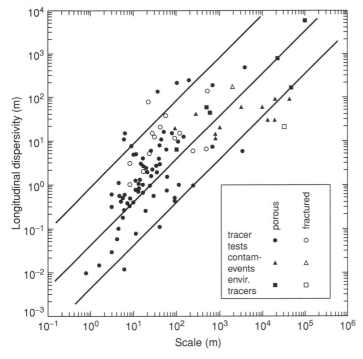

Figure 6.16. Field data on dispersion coefficients taken from Gelhar et al. (1992) and equation (6.57).

Our solution is similar to Example 2.3, except that we have a convective flux, a porosity, and

$$D \Rightarrow D_x, D_y, D_z$$

Then,

$$C = \frac{M}{8\varepsilon \left(\pi \, t/R\right)^{3/2} \sqrt{D_x D_y D_z}} \exp\left[-\frac{R\left(x - \frac{U}{R}t\right)^2}{4D_x t} - \frac{Ry^2}{4D_y t} - \frac{Rz^2}{4D_y t} \right] \quad (E6.12.1)$$

Now estimate the maximum concentration location of this maximum and spread of the atrazine cloud as a function of time.

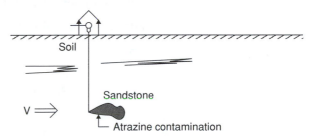

Figure E6.12.1. Illustration of a well and atrazine spill.

Dispersion Coefficients. As a first guess, let's use the empirical relations provided in equation (6.55) and (6.56). We will assume no heterogeneity in rock porosity and no lenses.

$$\frac{D_x}{D} = \frac{D_L}{D} \cong Pe = \frac{Ud}{D} \quad \text{or} \quad D_x = (10^{-4}\text{m/s})(10^{-3}\text{m}) = 10^{-7}\text{m}^2/\text{s}$$

$$\frac{D_y}{D} = \frac{D_z}{D} \cong 0.1 \; Pe \qquad \text{or} \quad D_y, \; D_z = 0.01 \, (10^{-7}\text{m}^2/\text{s}) = 10^{-8}\text{m}^2/\text{s}$$

Retardation Coefficient. The *Handbook of Fate and Exposure Data for Organic Chemicals, Vol. III: Pesticides* (published by CRC Press) gives

$$\log(K_{ow}) = 2.75 \quad \text{for atrazine}$$

$$\therefore K_{ow} = 10^{2.75} = 562$$

Using Karikhoff et al.'s (1979) relationship,

$$K_d = \beta f \, K_{ow} \quad \text{where} \quad \beta = 0.41 \, \text{cm}^3/\text{g} \quad \text{and} \quad f = 10^{-4}$$

Then

$$K_d = 0.41 \, \text{cm}^3/\text{g} \, (10^{-4})(562) = 0.023 \, \text{cm}^3/\text{g}$$

and

$$R = 1 + \frac{\rho_b}{\varepsilon} K_d = 1 + 6\frac{g}{\text{cm}^3}(0.023 \, \text{cm}^3/\text{g}) = 1.14$$

Even with a fairly sorptive organic compound, the retardation coefficient in rock is not much different from 1.0. Equation (E6.12.1) gives a maximum at $y = 0$, $z = 0$, and $x = Ut/R$. Thus,

$$C_{\max} = \frac{M}{8\varepsilon(\pi t/R)^{3/2}\sqrt{D_x D_y D_z}} \tag{E6.12.2}$$

and

$$X_{\max} = \frac{Ut}{R} = 0.88 \times 10^{-4} \, (\text{m/s}) \; t(\text{s}) \tag{E6.12.3}$$

We will indicate spread by 4σ, which corresponds to 95% of the total mass of the cloud for a Gaussian distribution like this equation provides.

$$\text{For one-dimensional diffusion:} \, \sigma^2 = 2 \, Dt/R$$

$$\text{For three-dimensional diffusion:} \, \sigma^2 = \frac{2}{9} Dt/R$$

and

$$4\sigma_x = \frac{4\sqrt{2D_x t/R}}{3} \tag{E6.12.4}$$

$$4\sigma_z = 4\sigma_y = \frac{4\sqrt{2D_y t/R}}{3} \tag{E6.12.5}$$

Table E6.12.1: *Estimated concentration over time and space with transport through a uniformly porous sandstone*

Time	C_{max} (g/m^3)	X_{max} (m)	$4\sigma_x$ (m)	$4\sigma_y$ (m)
1 hr	4.0×10^4	0.32	0.04	0.014
1 day	340	7.6	0.17	0.054
1 month	2.11	229	0.90	0. 29
1 yr	0.050	2,800	3.1	0.98
2 yrs	0.018	5,500	4.4	1.39
10 yrs	0.0016	27,000	9.9	3.13

Note that in Table E6.12.1, the concentrations at 1 hr and 1 day are above the initial atrazine concentration (1,000 g/m^3). This is one problem with Dirac delta boundary conditions because they initially have no volume, only mass. At greater elapsed time, however, the inaccuracies of the Dirac delta solution have a minimal impact on the resulting concentration.

However, virtually all media have heterogeneities of high- and low-porosity regions, as well as lenses that form around cracks. If we assume that the spacing of these regions is a mean of 1 m, instead of the 1 mm grain size, then our estimated dispersion coefficients are increased by a factor of 10^3. Applying equations (E6.12.2) through (E6.12.4) results in a reduction in C_{max} by a factor of 3.2×10^4 and an increase in both 4σ values by a factor of 32. The resulting estimates of the pertinent parameters are given in Table E6.12.2. These values are likely to be more realistic for transport through a groundwater aquifer.

The unknown dispersion coefficient is not uncommon in groundwater transport problems. It is typically one of the parameters fitted to measurements in groundwater transport.

EXAMPLE 6.13: *Drinking water contamination by trichloroethylene (steady-state groundwater transport with lateral dispersion)*

A military ammunition plant in Arden Hills, Minnesota, used trichloroethylene (TCE) as a metal cleaning solvent for many years. TCE is currently believed to be

Table E6.12.2: *Estimated concentration over time and space with transport through a sandstone media with 1 m heterogeneities*

Time	C_{max} (g/m^3)	X_{max} (m)	$4\sigma_x$ (m)	$4\sigma_y$ (m)
1 hr	1.25	0.32	1.3	0.4
1 day	0.011	7.6	5.5	1.7
1 month	6.6×10^{-5}	229	29	9.2
1 yr	1.57×10^{-6}	2,800	99	31
2 yrs	5.6×10^{-7}	5,500	140	44
10 yrs	5.0×10^{-8}	27,000	320	99

a carcinogen. Unaware of the hazardous nature of TCE, plant personnel placed the waste grease and TCE in a trench to burn (the grease) or soak into the ground (the TCE) and disappear from sight for many years. What was not known, however, is that they were placing the TCE into an aquifer that surfaces near the armory. Four kilometers downstream, the city of New Brighton used this aquifer as a source of municipal water supply. What is the expected TCE concentration in the New Brighton water supply and what should be done in the adjacent cities? Is the TCE plume sufficiently captured by the New Brighton wells?

The following conditions were approximated from available data:

Supply of TCE $= 8\,\text{kg/day}$
$U = 1.6 \times 10^{-5}\,\text{m/s}$
Aquifer thickness, $H = 30\,\text{m}$
$R \cong 1.0$ in the aquifer for TCE
Aquifer porosity, $\varepsilon = 0.3$
New Brighton extraction, $Q = 0.25\,\text{m}^3/\text{s}$
Drinking water recommended limit for TCE $= 5\mu\text{g/L}\,(5 \times 10^{-3}\text{g/m}^3)$

As a first assumption, we will assume that the New Brighton well was located in the center of the plume. Compute the capture zone and then the concentrations within this capture zone. The capture zone is given by

$$W = \frac{Q}{\varepsilon HU} = 1{,}740\,\text{m} \tag{E6.13.1}$$

At a velocity of $1.6 \times 10^{-5}\,\text{m/s}$, the 4-km distance would be covered in 8 years, which is short compared with the ~40 years of dumping TCE. We will therefore assume that the system is at steady state. The solution is similar to Example 5.3, except with dispersion in the longitudinal and lateral directions, and without an image. We have these boundary conditions:

1. At $x = 0$, $\dot{M} = 100\,\text{kg/day} = 1.2\,\text{g/s}$
2. At $y = \infty$, $C = 0$
3. At $x = \infty$, $C = 0$

Let's start with equation (E5.3.5):

$$\hat{C} = \frac{2\dot{M}}{h(4\pi\,\bar{\varepsilon}_y\,xU)^{1/2}}\left\{\exp\left(\frac{-Uy^2}{4\bar{\varepsilon}_y\,x}\right) + \left[\frac{-U(y - 2b)^2}{4\bar{\varepsilon}_y\,x}\right]\right\} \tag{E5.3.5}$$

and make these changes to meet our boundary conditions:

1. $\hat{C} \Rightarrow C$. There is no need to subtract a background concentration.
2. Convert to transverse dispersion coefficient, D_t, and eliminate the image:

$$\left\{\exp\left(\frac{-Uy^2}{4\bar{\varepsilon}_y\,x}\right) + \exp\left[\frac{-U(y - 2b)^2}{4\bar{\varepsilon}_y\,x}\right]\right\} \Rightarrow \exp\left(\frac{-Uy^2}{4D_t x}\right)$$

and

$$2\dot{M} = \dot{M}/\varepsilon$$

The end result is

$$C = \frac{\dot{M}}{h\varepsilon(4\pi x U D_t)^{1/2}} \exp\left(\frac{-Uy^2}{4D_t x}\right) \qquad \text{(E6.13.2)}$$

with a capture zone mean of

$$\overline{C} = \frac{2}{Y}\int_0^{Y/2} C\,dy \qquad \text{(E6.13.3)}$$

where Y is the width of the capture zone, and leakage from the capture zone, M_L, of

$$M_L = 2\varepsilon H \int_{Y/2}^{\infty} C\,dy \qquad \text{(E6.13.4)}$$

Using the rough approximations of equations (6.57) and (6.56), we get

$$D_L = 0.0026\,\text{m}^2/\text{s}$$

and

$$D_t = 2.6 \times 10^{-4}\,\text{m}^2/\text{s}.$$

Note that equation (6.57) gives $D_L = 0.0026\,\text{m}^2/\text{s}$. with a 67% confidence interval of between 0.064 and $1 \times 10^{-4}\,\text{m}^2/\text{s}$ or 1.4 orders of magnitude. This variation would need to be considered in any preliminary analysis of this problem.

The solution to equation (E6.13.2) for this application is given in Figure E6.13.1 at various lateral distances from the peak concentration. The capture zone mean is $0.42\,\text{g/m}^3$ or almost 100 times the recommended limit, which would raise concern in New Brighton. The leakage from the capture zone at $x = 4,000\,\text{m}$ for this scenario is computed to be 47 g, or sufficient mass to result in a concentration of $9 \times 10^{-4}\,\text{g/m}^3$ for a similar capture zone. It is possible then that almost all of the plume was captured by the city of New Brighton, so the problem may not cover a wider area than the immediate downstream cities. This, at least, is one positive result of the low transverse dispersion of groundwater plumes.

3. Dispersion in Rivers

Rivers are close to the perfect environmental flow for describing the flow as plug flow with dispersion. The flow is confined in the transverse and vertical directions, such that a cross-sectional mean velocity and concentration can be easily defined. In addition, there is less variation in rivers than there is, for example, in estuaries or reactors – both of which are also described by the plug flow with dispersion model. For that reason, the numerous tracer tests that have been made in rivers are useful to characterize longitudinal dispersion coefficient for use in untested river reaches. A sampling of the dispersion coefficients at various river reaches that were

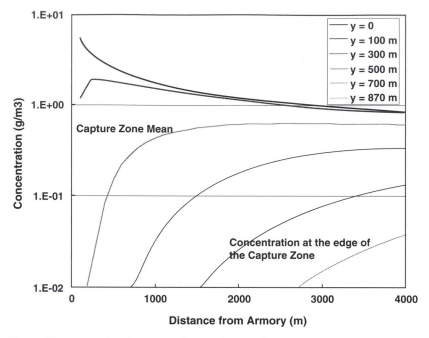

Figure E6.13.1. Predicted concentration of trichloroethylene in groundwater plume versus distance for the ammunitions plant release (city of New Brighton case study).

determined from tracer tests are given in Table 6.4. Also given are the relevant mean parameters for each reach.

The question that we need to ask ourselves is whether the longitudinal dispersion can be predicted accurately for these rivers. Equation (6.35), which predicts that $D_L/(u_* h) = $ constant, is shown in Table 6.4 to have a large range of constants, probably because of the variations in cross section and morphology seen in natural streams. Fisher (1973) observed that this constant seemed to depend on mean surface width, W, and substituted W for h in the numerator of equation (6.34) to develop the following empirical equation to characterize longitudinal dispersion coefficient in rivers:

$$D_{Lp} = 0.011 \frac{U^2 W^2}{u_* h} \qquad (6.58)$$

where D_{Lp} is the predicted value of D_L, h is the mean depth, U is the cross-sectional mean velocity, and u_* is the mean shear velocity of the river reach. When compared with the data given in Table 6.4, the root-mean-square error relative to the measurement, given by the equation,

$$\text{Relative rms errors} = \sqrt{\frac{1}{n} \sum_n \left(1 - \frac{D_{LP}}{D_L}\right)} \qquad (6.59)$$

Table 6.4: *Measurements of longitudinal dispersion coefficient in laboratory flumes and rivers (after Fisher et al., 1979)*

References	Channel	Depth. K (m)	Width. W (m)	Mean velocity, U (m/sec)	Shear velocity, u^* (m/sec)	Observed dispersion coefficient, D_L (m²/sec)	$D_{Lh} u^*$	D_{LR} predicted by Eq. (6.58) (m²/sec)
Thomas (1958)	Chicago Ship Canal	8.07	48.8	0.27	0.0191	3.0	20	12.4
State of California (1962)	Sacramento River	4.00		0.53	0.051	15	74	
Owens et al. (1964)	River Derwent	0.25		0.38	0.14	4.6	131	
Glover (1964)	South Platte River	0.46		0.66	0.069	16.2	510	
Schuster (1965)	Yuma Mesa A Canal	3.45		0.68	0.345	0.76	8.6	
Fischer (1967)	Trapezoidal laboratory channel with roughened sides	0.035	0.40	0.25	0.0202	0.123	174	0.156
		0.047	0.43	0.45	0.0359	0.253	150	0.244
		0.035	0.40	0.45	0.0351	0.415	338	0.298
		0.035	0.34	0.44	0.0348	0.250	205	0.202
		0.021	0.33	0.45	0.0328	0.400	392	0.248
		0.021	0.19	0.46	0.0388	0.220	270	0.103
Fischer (1968b)	Green-Duwamish River, Washington	1.10	20		0.049	6.5–8.5	120–160	
Yotsukura et al. (1970)	Missouri River	2.70	200	1.55	0.074	1500	7500	3440
Godfrey and Frederick (1970)	Copper Creek, Virginia (below gauge)	0.49	16	0.27	0.080	20	500	5.24
		0.85	18	0.60	0.100	21	250	15.1
		0.49	16	0.26	0.080	9.5	245	4.86
	Clinch River, Tennessee	0.85	47	0.32	0.067	14	235	22
		2.10	60	0.94	0.104	54	245	73
		2.10	53	0.83	0.107	47	210	28
	Copper Creek, Virginia (above gauge)	0.40	19	0.16	0.116	9.9	220	2.19
	Powell River, Tennessee	0.85	34	0.15	0.055	9.5	200	6.12
	Cinch River, Virginia	0.58	36	0.21	0.049	8.1	280	22.1
	Coachella Canal, California	1.56	24	0.71	0.043	9.6	140	47.6
McQuivey and Keefer (1974)	Bayou Anacoco	0.94	26	0.34	0.067	33	520	13
		0.91	37	0.40	0.067	39	690	38
	Nooksack River	0.76	64	0.67	0.27	35	170	98
	Wind Bighorn Rivers	1.10	59	0.88	0.12	42	330	232
		2.16	69	1.55	0.17	160	440	340
	John Day River	0.58	25	1.01	0.14	14	170	88
		2.47	34	0.82	0.18	65	150	20
	Comite River	0.43	16	0.37	0.05	14	650	16
	Sabine River	2.04	104	0.58	0.05	315	3090	330
		4.75	127	0.64	0.08	670	1760	190
	Yadkin River	2.35	70	0.43	0.10	110	470	44
		3.84	72	0.76	0.13	260	520	68
Gulliver (1977)	MERS Experimental Streams	0.32	32	0.95	0.52	0.11	17	0.61

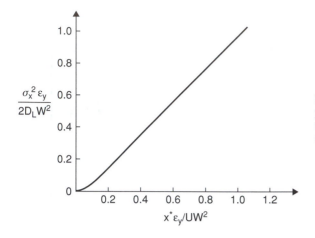

Figure 6.17. Growth of tracer variance following a pulse injection. The analysis should begin downstream of $x^* = 0.2$. Adapted from Fisher et al. (1979).

where n is the number of measurements is 1.71. This relative rms error means that roughly 67% of the predictions are within a factor of 1.71 (within 71%) of the observed.

Tracer Determination of Longitudinal Dispersion Coefficient in Rivers. Tracers are generally used to determine longitudinal dispersion coefficient in rivers. Some distance is required, however, before the lateral turbulent diffusion is balanced by longitudinal convection, similar to Taylor's (1953) analysis of dispersion in a laminar flow. This transport balancing distance, x^* is given by the equation

$$x^* = \frac{0.2UW^2}{\varepsilon_y} \tag{6.60}$$

The region $x < x^*$ can be visualized as a mixing region, which can skew the results of a tracer study. Downstream of this region, where turbulent diffusion is balanced by longitudinal convection, the variance of a tracer pulse grows linearly with distance, as shown in Figure 6.17. It is best to begin the measurements a distance x^* below the tracer release. The technique used to perform the analysis of tracer studies will be the subject of Example 6.14.

EXAMPLE 6.14: *Determination of longitudinal dispersion coefficient in a river.*

You are part of a forensic investigation of an oil spill into the Nemadji River, Wisconsin. A tanker car carrying a solvent derailed on a bridge and fell into the river. The forensic investigation team does have a computational model that will simulate the spill, if some coefficients are determined, including D_L and t_r. You have decided that the most cost-effective means of determining these parameters would be to perform a conservative tracer pulse test and adjust the parameters from discharge on the day of the tracer test (3 m³/s) to discharge on the day of the spill (8 m³/s) with some equations that have been developed.

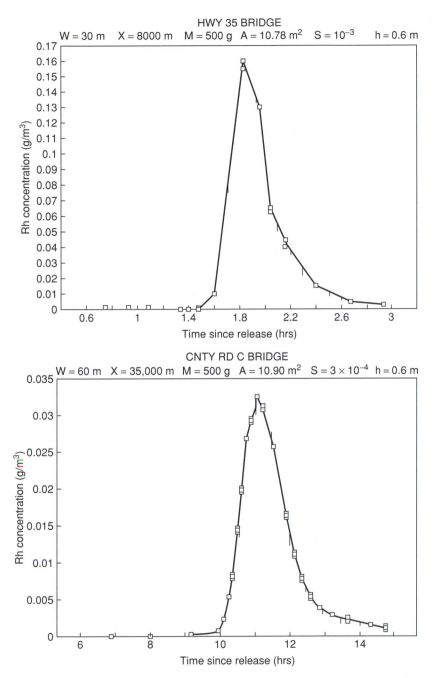

Figure E6.14.1. Tracer measurements taken at the Highway 35 and County Road C bridges.

The location of the pulse input is identified as $x = 0$. The measured tracer concentrations and other relevant data are given in Figure E6.14.1. From this data, determine the D_L and t_r parameters on the day of the test for the reach from $x = 8,000$ m to $x = 35,000$ m.

Because the variance of the tracer curve grows linearly with distance after $x = x^*$, we can make the following statements:

$$\Delta t_r = t_{r_2} - t_{r_1} \tag{E6.14.1}$$

and

$$\Delta \sigma_t^2 = \sigma_t^2|_2 - \sigma_t^2|_1 \tag{E6.14.2}$$

and, finally,

$$\Delta \sigma^2 = \frac{\Delta \sigma_t^2}{\Delta t_r^2} = \frac{2 D_L}{U \Delta x} = \frac{2 D_L \, \Delta t_r}{\Delta x^2} \tag{E6.14.3}$$

At $x = 8,000$ m,

$$t_{r_1} = \frac{\int Ct \, dt}{\int C \, dt} \cong \frac{\sum Ct \, \Delta t}{\sum C \, \Delta t} \tag{E6.14.4}$$

Using all of the Δt values set equal to 0.1 hr, Δt can be cancelled out of the equation, and

$$t_{r_1} = \frac{1.30 \text{ hrs g/m}^3}{0.67 \text{ g/m}^3} = 1.94 \text{ hrs} \tag{E6.14.5}$$

Also,

$$\sigma_t^2 \Big|_1 = \frac{\int Ct^2 \, dt}{\int C \, dt} - t_r^2 \cong \frac{\sum Ct^2 \, \Delta t}{\sum C \, \Delta t} \tag{E6.14.6}$$

or, again dropping the equal Δt's:

$$\sigma_t^2 \Big|_1 = \frac{2.59 \text{ hrs}^2 \text{g/m}^3}{0.666 \text{ g/m}^3} - (1.94 \text{ hrs})^2 = 0.13 \text{ hr}^2 \tag{E6.14.7}$$

We have assumed that the location 8,000 m downstream of the pulse injection will be out of the "mixing" region, as specified by equation (6.60). We are now ready to check this assumption with the parameters of the stream flow at $x = 8,000$ m. First $U = Q/A = 8 \text{ m}^3/\text{s}/10.8 \text{ m} = 0.74 \text{ m/s}$, and from Figure 4.7d:

$$\varepsilon_y \cong (0.6 \pm 0.3) u_* h \tag{E6.14.8}$$

where

$$u_* = \sqrt{ghS} = \sqrt{(9.8 \text{ m/s}^2)(0.6 \text{ m})(10^{-3})} = 0.077 \text{ m/s} \tag{E6.14.9}$$

Then,

$$\varepsilon_y = (0.6 \pm 0.3)(0.077 \text{ m/s})(0.6 \text{ m}) = 0.028 \pm 0.014 \text{ m}^2/\text{s} \tag{E6.14.10}$$

and equation (6.52) gives

$$x^* = \frac{0.2(0.74 \text{ m/s})(30 \text{ m})^2}{0.028 \text{ m}^2/\text{s}} = 4,760 \text{ m} \tag{E6.14.11}$$

With the approximations in equation (6.60), $x = 8,000$ m will be assumed sufficient to begin our determination of D_L, especially because that is one of the few access points (a bridge) into this reach of the river.

We will now perform a similar calculation on the tracer cloud at $x = 35,000$ m.

$$t_{r_2} = 11.54 \text{ hrs} \tag{E6.14.12}$$

$$\sigma_t^2 \Big|_2 = \frac{14 \text{ hrs}^2 \text{ g/m}^3}{0.105 \text{ g/m}^3} - (11.54 \text{ hrs})^2 = 0.837 \text{ hr}^2 \tag{E6.14.13}$$

$$\Delta x = 35,000 \text{ m} - 8,000 \text{ m} = 27,000 \text{ m} \tag{E6.14.14}$$

Now, rearranging equation (E6.14.3),

$$D_L = \frac{\Delta \sigma_t^2 \Delta x^2}{2 \Delta t_r^3} = \frac{(0.837 - 0.13) \text{ hr}^2 (27,000 \text{ m})^2}{2(11.54 - 1.94)^3 \text{ hrs}^3} \tag{E6.14.15}$$

or

$$D_L = 291,000 \text{ m}^2/\text{hr} = 81 \text{ m}^2/\text{s} \tag{E6.14.16}$$

We can use equation (6.58) to adjust our dispersion coefficient from the 8 m³/s with 0.6 m mean depth on the day of the tracer test to the 3 m³/s with 0.4 m mean depth that existed as the river discharge on the day of the spill:

$$D_L = 0.011 \, (Q/Wh)^2 \, W^2/(\sqrt{ghS}\,h) \tag{E6.14.17}$$

and then assuming that the slope does not change and that the banks are fairly steep, such that $dA = W\,dh$, and assigning the subscripts t and s to indicate tracer and spill:

$$D_{LS} = D_{Lt} \left(\frac{Q_s}{Q_t}\right)^2 \left(\frac{h_t}{h_s}\right)^{7/2} = 81 \frac{\text{m}^2}{\text{s}} \left(\frac{3 \text{ m}^3/\text{s}}{8 \text{ m}^3/\text{s}}\right)^2 \left(\frac{0.6 \text{ m}}{0.4 \text{ m}}\right)^{7/2} = 47 \text{ m}^2/\text{s} \tag{E6.14.18}$$

In addition, t_r can be adjusted as well:

$$\Delta t_{rs} = \Delta t_{rt} \frac{Q_t}{Q_s} \frac{h_s}{h_t} = (11.54 - 1.94) \text{ hrs} \frac{8 \text{ m}^3/\text{s}}{3 \text{ m}^3/\text{s}} \frac{0.4 \text{ m}}{0.6 \text{ m}} \tag{E6.14.19}$$

so

$$\Delta t_{rs} = 17 \text{ hrs} \tag{E6.14.20}$$

The use of an empirical relation, such as equation (6.58), to adjust parameters for discharge is more accurate than simply using the equation itself, because the coefficient and other variables that drop out of the equation can have significant error.

G. Problems

1. Lake Ontario contains 1,600 km^3 of water, with discharge into the lake of 5,500 m^3/s from the Niagara River discharge out of the lake of 6,600 m^3/s, precipitation of 0.9 m/yr and evaporation of 0.8 m/yr. Tributaries account for the remainder. In 1980, phosphorus removal was required of all municipal water treatment plants in the Great Lakes States to reduce the algal population of these lakes. Assume that Lake Ontario may be represented by five complete mix reactors. Most of the phosphorus would be tied up in phytoplankton, which has a net settling velocity of 8 m/yr in the surface layer and 16 m/yr in the lower layer where there is little light and the phytoplankton are dying.

   a. Compute the steady-state total phosphorus concentrations in the lake before 1980.

   b. Compute the steady-state total phosphorus concentrations after municipal treatment began in 1980.

   c. What length of time is required before the upper and lower layers of the lake reach within 10% of a new steady state after phosphorus reduction? How does this compare with the residence time of the lake?

   Note: t(within 10%) occurs at $\dfrac{C(t) - C(\text{part b})}{C(\text{part a}) - C(\text{part b})} = 0.10$

   Relevant data:
   Lake surface area: 18,800 km^2
   Mean surface layer depth: 10 m
   Exchange between reactors: ~1,000 m/s
   Inflow P concentration Niagara River $= C(\mu g/L) = 12 + 14\,e^{-4t}$ (where t is in years from 1980)

Phosphorus concentration	Before 1980	After 1980
From tributaries	82 μg/L	30 μg/L
Phosphorus loading from municipal discharge	3.5 × 10^6 kg/yr	0.5 × 10^6 kg/yr
Atmospheric loading	1.7 × 10^6 kg/yr	1.7 × 10^6 kg/yr

2. Model the reach given in Figure E6.14.1 (Highway 35 bridge) with a set of ideal reactor elements to match this conservative tracer data. Plot the data and model prediction together.

3. If the aforementioned river was modeled as a plug flow reactor with dispersion, what would be the dispersion coefficient and the cross-sectional mean velocity? How does the predicted curve compare with those plotted for problem 2 (i.e., plot it)?

4. An ideal reactor model is composed of complete mix reactors with leaky dead zones that are also complete mixed. Assuming a pulse input to reactor 1, derive the concentration versus time equation for the third reactor–dead zone combination. Derive the concentration versus time equation for the nth reactor–dead zone combination.

5. You have placed an air diffuser (sparger) at the bottom of a 20-m-deep lake for the purpose of increasing oxygen concentrations. The primary cost of spargers is the cost of pressurizing the air to overcome hydrostatic pressure and head losses. The bubbling action adds oxygen to the water through air bubble–water transfer and also through surface air–water transfer. A steady-state water concentration is reached. Is this steady concentration at equilibrium? Why?

6. A burning dump emits 5 g/s of nitrous oxide (NO), spread over 50 m × 50 m, with an initial cloud height of 3 m. Estimate the maximum downwind concentration of NO at 5 km from the dump when the wind speed is 8 m/s at 3 m height over farmland, with a roughness length of 0.2 m. The threshold limit for NO is 25 ppm(v). Is this exceeded at 5 km from the dump?

7. The following data were collected on a conservative dye ($k = 0$) in the Prairie River and St. Lawrence River in the vicinity of Montreal. A discharge of the Prairie River just before its entrance into the St. Lawrence River was 50 m^3/s with the dye concentration of 5 mg/L, and the discharge of the St. Lawrence River was 100 m^3/s with the dye concentration of 10 mg/L.

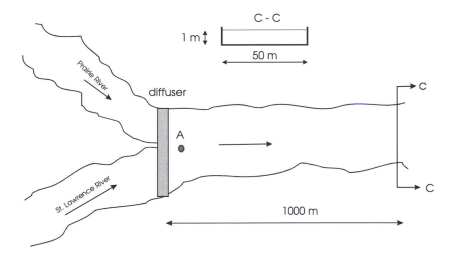

a. What is the concentration at point A behind the diffuser? (Assume complete mixing between the two rivers.)

b. What is total mass (kg) of the dye in the channel if the measured peak concentration is 4 mg/L at the point 1,000 m downstream from the diffuser? The longitudinal dispersion coefficient is $0.5 \text{ m}^2/\text{s}$. Assume that peak concentration is caused by the instantaneous (impulse) dye injection into the channel.

c. Explain and provide a schematic for the equilibrium mixing approximation in open-channel flow.

7 Computational Mass Transport

This chapter primarily deals with analytical solutions to the diffusion equation, where all that is needed is a piece of paper, something to write with, and knowledge of the solution techniques that are applicable. There are, however, many boundary conditions that cannot be well simulated with the simple boundary conditions used herein and require a digital solution with a computer. The numerical integrals given in equations (2.49) and (2.50) and in Example 2.9 were a simple form of these digital solutions, although they would be classified as a numerical integration.

Computational mass transport involves (1) discretization (division into discrete elements) of the spatial domain into control volumes with an assumed equal concentration, similar to complete mixed tanks; (2) discretization of time into steps of Δt; and (3) computing the fluxes into and out of each control volume over time to determine the concentration in each control volume over time. Once we have the discretization completed, all we will be doing is adding, subtracting, multiplying, and dividing numbers. Of course, for our computational solution to approach the real solution, the control volumes must be sufficiently small and the time steps must be sufficiently short. We will therefore be doing a great deal of adding, subtracting, multiplying, and dividing and would welcome the assistance of a computer.

The analytical solutions are still useful when moving to more realistic boundary conditions and a computational solution. Regardless of how often a computational model has been run, there are discretization variables (such as Δx, Δy, Δz, and Δt) in each application that need to be determined. Typically, in computational techniques, the distance or time step is reduced to get a feel for the magnitude of the error generated by discretization into finite control volumes. The concept is that greater than half of the error will be eliminated by dividing the step size by 2. This will often give an indication, but is not sufficient to determine if the computational routine is accurate, and there are cases in which the indication will be wrong.

The only way to know if your computational model is functioning properly is to test it against an analytical solution. This is done by first choosing an analytic solution that has boundary conditions close to those that will be modeled computationally.

Then run the computational model on the same conditions as the analytical solution was applied. The result will be a direct, apples-to-apples comparison that will let you know how your computational solution is performing in the planned application. Then, you are ready to move on to the more complex boundary conditions of the application.

A. Computational Terminology

The terminology of computational techniques is descriptive, but one needs to know what is being described. Table 7.1 lists some common terms with a definition relative to mass transport. Most computational techniques in fluid transport are described with control volume elements, wherein the important process to be computed is the transport across the interfaces of small control volumes. The common control volumes are cubes, cylindrical shells, triangular prisms, and trapezoidal prisms, although any shape can be used. We will present the control volume technique.

Table 7.1: *Selective terminology of computational techniques*

Control volume method	Designed for conditions with fluxes across interfaces of small, well-mixed elements – primarily used in fluid transport
Finite element method	Extrapolates parameters between nodes. Predominant in the analysis of solids, and sometimes used in groundwater flow.
Boundary element method and analytic element method	Functions with Laplace's equation, which describes highly viscous flow, such as in groundwater, and inviscid flow, which occurs far from boundaries.
Distance discretization:	
Upwind differences	Flux between control volumes is determined by conditions in the upwind control volume.
Central differences	Flux between control volumes is determined by the average condition of the two control volumes.
Exponential differences	Flux between control volumes is given by a weighted mean of conditions in both control volumes, as determined by the one-dimensional, steady flux solution.
Time discretization:	
Explicit	Conditions in the previous time step are used to predict the change in conditions for the next time step.
Implicit (Crank–Nicolson)	An average of conditions in the previous and next time step are used to predict the change in conditions for the next time step. Often requires iteration.
Fully implicit (Laasonen)	Conditions in the next time step are used to predict the change in conditions for the next time step. Often requires iteration.

Table 7.1 has three selections with regard to distance discretization: upwind, central, and exponential differences. Upwind differences tend to be more stable and are more accurate than central differences when dealing with advective fluxes that are dominant over diffusive fluxes. Central differences are designed for dominant diffusive fluxes, but the computational solution can develop an oscillatory instability when large concentration gradients are encountered. Exponential differences utilize the steady-state, one-dimensional solution for flux at the interface between two locations to determine an intermediate weighting between upwind and central differences. These are typically more stable and accurate than either central or upwind differences. At relatively large advective flux, the exponential differences become upwind differences. At large diffusive flux, they become central differences.

The choice of time discretization is determined by preference and the complexity of the equation to be solved. Explicit differences are simple and do not require iteration, but they can require small time steps for an accurate solution. Implicit and fully implicit differences allow for accurate solutions at significantly larger time steps but require additional computations through the iterations that are normally required.

We will begin with domain discretization into control volumes. Consider our box used in Chapter 2 to derive the mass transport equation. Now, assume that this box does not become infinitely small and is a control volume of dimensions Δx, Δy, and Δz. A similar operation on the entire domain, shown in Figure 7.1, will discretize the domain into control volumes of boxes. Each box is identified by an integer (i, j, k), corresponding to the box number in the x-, y-, and z-coordinate system. Our differential domain has become a discrete domain, with each box acting as a complete mixed tank. Then, we will apply our general mass conservation equation from Chapter 2:

$$\underset{\text{IN}}{\text{Flux rate}} - \underset{\text{OUT}}{\text{Flux rate}} + \underset{\text{rate}}{\text{Source}} - \underset{\text{rate}}{\text{Sink}} = \text{Accumulation} \tag{2.1}$$

We could perform our discretization on the governing partial differential equations, which are different for each case, but there is no need to go beyond the most fundamental equation of mass transport, equation (2.1).

B. Explicit, Central Difference Solutions

Central differences are typically used when the flux is predominately due to a diffusive term, where the gradients on both sides of the control volume are important. An explicit equation is one where the unknown variable can be isolated on one side of the equation. To get an explicit equation in a computational routine, we must use the flux and source/sink quantities of the previous time step to predict the concentration of the next time step.

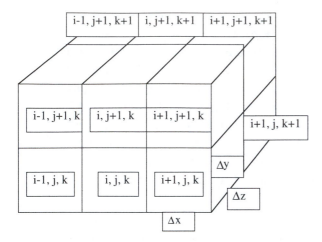

Figure 7.1. Discretized flow and mass transport domain in Cartesian coordinates.

We will use the diffusive flux term of equation (2.2), applied at the $i+1/2$ interface between the (i, j, k) control volume and the $(i + 1, j, k)$ control volume:

$$\text{Diffusive flux rate}|_{i+1/2} = -D_{i+1/2}\left(\frac{\partial C}{\partial x}\right)_{i+1/2} \Delta y \Delta z \qquad (7.1)$$

where

$$\left(\frac{\partial C}{\partial x}\right)_{i+1/2} \equiv \frac{C_{i+1} - C_i}{\Delta x} \qquad (7.2)$$

and $D_{i+1/2}$ is the harmonic mean of D_i and D_{i+1}, similar to resistances in series (e.g., a zero diffusion coefficient should result in no diffusion):

$$\frac{1}{D_{i+1/2}} = \frac{1}{2D_{i+1}} + \frac{1}{2D_i} \qquad (7.3a)$$

or

$$D_{i+1/2} = \frac{2D_i D_{i+1}}{D_i + D_{i+1}} \qquad (7.3b)$$

Then equation (7.1) becomes, at the (k, j) location,

$$\text{Diffusive flux rate}|_{i+1/2} = -\frac{2D_i D_{i+1}}{D_i + D_{i+1}} \frac{C_{i+1} - C_i}{\Delta x} \Delta y \Delta z \qquad (7.4)$$

Applying the same logic to the $i - 1/2$ interface results in

$$\text{Diffusive flux rate}|_{i-1/2} = -\frac{2D_i D_{i-1}}{D_i + D_{i-1}} \frac{C_i - C_{i-1}}{\Delta x} \Delta y \Delta z \qquad (7.5)$$

We will apply the computational solution to some of the examples that have already been solved analytically and compare the solutions.

EXAMPLE 7.1: *Unsteady dissolution of ammonia into groundwater (unsteady, one-dimensional solution with pulse boundary conditions) solved with an explicit, central difference computation*

Solve the tanker truck spill problem of Example 2.2 using explicit, central differences to predict concentrations over time in the groundwater table. Compare these with those of the analytical solution. The mass spilled is 3,000 kg of ammonia over 100 m^2, and the effective dispersion coefficient through the groundwater matrix is 10^{-6} m^2/s.

Assume:

1. Minimal horizontal variations

$$0 \cong \frac{\partial C}{\partial x} \cong \frac{\partial C}{\partial y} \tag{7.6}$$

2. No flow in the vertical direction, $w = 0$
3. No reactions, including adsorption and desorption, such that $S = 0$

We will simulate the initial conditions with these boundary conditions:

1. The mass of chemical is assumed to be spread instantaneously across a very thin layer at $t = 0$ (a Dirac delta in z and t). At $z = 0^{+}$, $t = 0$, the total mass $=$ M; and the total surface area is A.
2. At $z \Rightarrow \infty$, $C \Rightarrow 0$

To discretize equation (2.1), we will identify elements along the z-coordinate as k and elements along the time coordinate as n. Then, the individual terms become

$$\text{Accumulation} \Rightarrow \frac{C_{k,n+1} - C_{k,n}}{\Delta t} \Delta x \Delta y \Delta z \tag{E7.1.1}$$

The flux terms are discretized at the cell interface. Because diffusion occurs in both directions, we will use central differences. In addition, we will use an explicit solution technique by discretizing our flux terms at only the n time step:

$$\left.\frac{\partial C}{\partial z}\right|_{k+1/2,n} \Rightarrow \frac{C_{k+1,n} - C_{k,n}}{\Delta z} \qquad \left.\frac{\partial C}{\partial z}\right|_{k-1/2,n} \Rightarrow \frac{C_{k,n} - C_{k-1,n}}{\Delta z}$$

In addition, since $D_k = D_{k+1} = D$, equations (7.3) and (7.4) become

$$D_{k+1/2} = D \quad \text{and} \quad D_{k-1/2} = D$$

Then,

$$\text{Flux in} - \text{Flux out} = D \left(\frac{C_{k+1,n} - C_{k,n}}{\Delta z} - \frac{C_{k,n} - C_{k-1,n}}{\Delta z} \right) \Delta x \Delta y$$

$$= D \frac{C_{k+1,n} - 2C_{k,n} + C_{k-1,n}}{\Delta z} \Delta x \Delta y \tag{E7.1.2}$$

Now, rewriting equation (2.1) using our discretization and dividing by $\Delta x \Delta y \Delta z$:

$$\frac{C_{k,n+1} - C_{k,n}}{\Delta t} = D\frac{C_{k+1,n} - 2C_{k,n} + C_{k-1,n}}{\Delta z^2} \tag{E7.1.3}$$

Equation (E7.1.3) is an approximation because we have discretized a smooth gradient into a series of well-mixed cells. If the cells are small (Δz is small) and the time step is small, there is little difference between the two equations. The question is, how small is small enough?

Given a distribution of oxygen at an initial time, n, through the boundary conditions, the concentration profile at time $n + 1$ can be found from rearranging equation (E7.1.3):

$$C_{k,n+1} = \frac{D\Delta t}{\Delta z^2}\left(C_{k+1,n} + C_{k-1,n}\right) + \left(1 - \frac{2D\Delta t}{\Delta z^2}\right)C_{k,n} \tag{E7.1.4}$$

Because the entire right-hand side of equation (E7.1.4) is known, $C_{k,n+1}$ can be found for the entire series of elements by applying equation (E7.1.4) sequentially for $k = 1$ at $z = 0$ through until $k = K$, where K is the last element of those in the discretized domain. Then, $n + 1$ becomes the previous time step, n, and the process continues until concentrations and fluxes are computed to the final time of interest.

This discretization of equation (2.1) uses control volumes with explicit, central differences. They are explicit because only the *accumulation* term contains a concentration at the $n + 1$ time step, resulting in an explicit equation for $C_{k,n+1}$ (equation (E7.1.4)). Another common option would be fully implicit (Laasonen) discretization where *flux rate* terms in equation (E7.1.2) are computed at the $n + 1$ time increment, instead of the n increment. This results in an equation for $C_{k,n+1}$ that is more difficult to solve but allows for larger time steps (Δt).

The dimensionless parameter, $D\Delta t/\Delta z^2$ is known as the diffusion number, Di. For stability of an explicit computational solution (i.e., for the solution to make any sense at all), this number must be less than $1/2$. The logic behind this stability criterion can be seen in equation (E7.1.4). If Di is greater than $1/2$, then the relation between $C_{k,n+1}$ and $C_{k,n}$ is negative, creating a computational solution that can have serious errors.

We will use a Δz of 0.1 m. Then, our maximum time step is

$$\Delta t \le \frac{\Delta z^2}{2D} = \frac{(0.1)^2}{2 \cdot 10^{-6}} = 5{,}000\,\text{s} = 1.4\,\text{hrs} \tag{E7.1.5}$$

We will select $\Delta t = 3{,}600$ s, which results in a diffusion number of 0.36. Our initial concentration in the surface control volumes is

$$C_{k,1} = \frac{M}{A \cdot \Delta z} = \frac{3 \times 10^6\,\text{g}}{100\,\text{m}^2 \cdot 0.1\,\text{m}} = 3 \times 10^5\,\text{g/m}^3 \tag{E7.1.6}$$

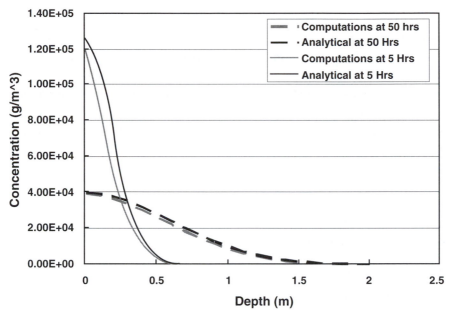

Figure E7.1.1. Comparison of the computational solution to the analytical solution of Example 2.2.

In Example 2.2, with boundary conditions 1 and 2, the analytical solution to this problem was found to be

$$C = \frac{2\,M/A}{\sqrt{4\pi\,Dt}}\,e^{-z^2/4Dt} \tag{E2.2.3}$$

The computational solution is compared with the analytical solution after 5 and 50 hrs following the spill in Figure E7.1.1. Because the stability criteria of $Di = 0.5$ is satisfied, the error cannot be reduced significantly by simply reducing the time step. It can be reduced, however, by reducing the distance step and staying within the stability criteria.

EXAMPLE 7.2: *Effect of fall turnover on oxygen concentration in lake sediments (unsteady, one-dimensional transport with step boundary conditions and a first-order sink, solved using explicit, central differences)*

Before fall turnover, there is zero oxygen concentration in the sediments and in the water above the sediments of Lake Harriet. At fall turnover, stratification of the lake is broken down and the water overlying the sediments abruptly reaches approximately C_0. You are interested in how fast the sediments will respond to the higher oxygen concentration. To determine, this you must answer two questions: (1) What is the oxygen profile in the sediments over time? (2) What is the flux rate across the sediment–water interface over time?

Given:

1. At $t = 0$, there is no oxygen in the sediments.
2. At $t = 0$, the water over the sediments suddenly has an oxygen concentration of C_0.
3. There is a first-order sink of oxygen in the sediments due to microbial degradation, such that $S = -k\,C$, where $k = 10^{-3}\,\text{s}^{-1}$.
4. The diffusion coefficient of oxygen in the sediments is 0.5 times the coefficient in water or $D = 1 \times 10^{-9}\,\text{m}^2/\text{s}$.

Assume:

1. No flow in the sediments (i.e., $u = v = w = 0$)
2. Variation in the x- and y-directions are small (i.e., $\partial C/\partial z \gg \partial C/\partial x$, $\partial C/\partial y$)
3. No sorption of oxygen to the sediments ($R = 1$)

With boundary conditions:

1. At $t \geq 0$, $z = 0$, $C = C_0$
2. At $t = 0$, $z > 0$, $C = 0$

To discretize equation (2.1), we will identify elements along the z-coordinate as k and elements along the time coordinate as n. Then, the individual terms become

$$\text{Accumulation} \Rightarrow \frac{C_{k,n+1} - C_{k,n}}{\Delta t} \Delta x \Delta y \Delta z \tag{E7.2.1}$$

The flux terms are discretized at the cell interface. Because diffusion occurs in both directions, we will use central differences. In addition, we will use an explicit solution technique by discretizing our flux terms at only the n time step:

$$\left.\frac{\partial C}{\partial z}\right|_{k+1/2,n} \Rightarrow \frac{C_{k+1,n} - C_{k,n}}{\Delta z} \qquad \left.\frac{\partial C}{\partial z}\right|_{k-1/2,n} \Rightarrow \frac{C_{k,n} - C_{k-1,n}}{\Delta z}$$

In addition, because $D_k = D_{k+1} = D$, equations (7.3) and (7.4) become

$$D_{k+1/2} = D \text{ and } D_{k-1/2} = D$$

Then,

$$\text{Flux in} - \text{Flux out} = D\left[\frac{C_{k+1,n} - C_{k,n}}{\Delta z} - \frac{C_{k,n} - C_{k-1,n}}{\Delta z}\right]\Delta x \Delta y$$

or

$$\text{Flux in} - \text{Flux out} = \frac{C_{k+1,n} - 2C_{k,n} + C_{k-1,n}}{\Delta z}\Delta x \Delta y \tag{E7.2.2}$$

and

$$\text{Source rate} - \text{Sink rate} = -kC_{k,n}\Delta x \Delta y \Delta z \tag{E7.2.3}$$

Now, rewriting equation (2.1) using our discretization and dividing by $\Delta x \Delta y \Delta z$:

$$\frac{C_{k,n+1} - C_{k,n}}{\Delta t} = D\frac{C_{k+1,n} - 2C_{k,n} + C_{k-1,n}}{\Delta z^2} - kC_{k,n} \tag{E7.2.4}$$

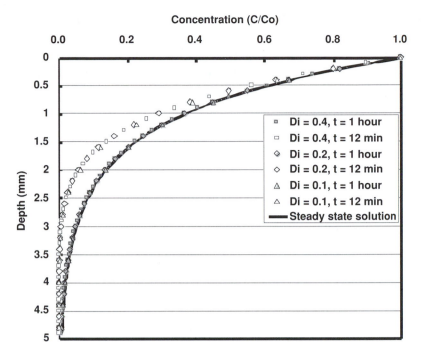

Figure E7.2.1. Computational solution for sediments suddenly exposed to oxygen in the water column. Also shown is the steady-state solution equation (E2.1.5).

Given a distribution of oxygen at an initial time, n, through the boundary conditions, the concentration profile at time $n + 1$ can be found from rearranging equation (E7.2.4):

$$C_{k,n+1} = Di \left(C_{k+1,n} + C_{k-1,n} \right) - \left(1 - 2Di - k\Delta t \right) C_{k,n} \qquad \text{(E7.2.5)}$$

The computational solution is given at 12 min and 1 hr in Figure E7.2.1, along with the steady-state solution from Example 2.1. As can be seen in Figure E7.2.1, the different values of Di did not greatly affect the computational solution. The solutions for $Di = 0.4, 0.2,$ and 0.1 are close to each other at 12 min and at 1 hr. In addition, after 1 hr, the solution is close to steady state. The reason is the relatively fast reaction involving oxygen. Without this sink, a steady-state solution would not exist. Considering the time scales of sediment oxygen demand, weeks and months, it is acceptable to consider the sediments at steady state, with an equation such as equation (E2.1.5). For this case, the computational routine is unnecessary.

C. Explicit, Upwind Difference Solutions

Upwind differences are typical for convective flux, where the upstream concentration is important to determine the convective flux at the upstream interface. Upwind differences have a lower "numerical diffusion" than central differences when applied to convective fluxes. In addition, upwind differences comes directly out of the mixed-cell concept (Section 6.C), and the tanks-in-series can be used to directly determine numerical diffusion (an apparent diffusion caused by the conversion of a continuous domain into a domain of discrete, well-mixed control volumes).

We will again consider the flux at the $i + 1/2$ interface between the (i, j, k) control volume and the $(i + 1, j, k)$ control volume given in Figure 7.1. The convective flux, however, is given with equation (2.4):

$$\text{Convective flux rate} = u\, C\, A_x \tag{2.4}$$

or

$$\text{Convective flux rate}_{i+1/2} = u_i\, C_i\, \Delta y\, \Delta z \tag{7.6}$$

Then, at the $(i - 1/2, j, k)$ interface:

$$\text{Convective flux rate}_{i-1/2} = u_{i-1} C_{i-1} \Delta y \Delta z \tag{7.7}$$

The diffusive fluxes are still appropriately defined with central differences. However, in two- and three-dimensional computational domains, where there are two and three independent variables, it is difficult to know in which direction the flow is moving at all locations and times. For example, in our one-dimensional case, if u is a negative value, then our flux terms at the interfaces would become

$$uC|_{i+1/2,n} \Rightarrow uC|_{i+1,n} \quad \text{and} \quad uC|_{i-1/2,n} \Rightarrow uC|_{i,n} \quad (u = -\text{value}) \tag{7.8}$$

We will therefore use a notation, $\| F^+, F^- \|$, which indicates the greater of F^+ and F^-. The fluxes at each interface become

$$uC|_{i+1/2,n} \Delta y \Delta z = \| u_{i,n} C_{i,n}, -u_{i+1,n} C_{i+1,n} \| \Delta y \Delta z \tag{7.9a}$$

and

$$uC|_{i-1/2,n} \Delta y \Delta z = \| u_{i-1,n} C_{i-1,n}, -u_{i,n} C_{i,n} \| \Delta y \Delta z \tag{7.9b}$$

Then, the concentration at the next time step, similar to equation (E7.2.4), would be given as

$$C_{i,n+1} = Cou_{i-1,n} \left\| 0, \frac{u}{|u|} C_{i-1,n} \right\| + (1 - Cou_{i,n})\, C_{i,n} + Cou_{i+1,n} \left\| 0, \frac{-u}{|u|} C_{i+1,n} \right\| \tag{7.10}$$

where $Cou = |u|\, \Delta t / \Delta x$ is the Courant number, which uses the magnitude of the velocity, u, and is thus always positive.

EXAMPLE 7.3: *Filter breakthrough with explicit, upwind differences*

A vertically oriented sand filter has multiple reactions occurring in the media, which cannot be modeled analytically. The flow in the filter is close to a plug flow. Determine the extent to which your convective routine – with explicit, upwind difference computations – can model the breakthrough of a concentration front moving through the filter, without reaction, so that you can later use it to simulate the flow with reactions. Mean flow velocity in the filter is 1 mm/s. Filter depth is 2 m.

We will again discretize equation (2.1), with elements along the z-coordinate as k and elements along the time coordinate as n. Then, the individual terms become

$$\text{Accumulation} \Rightarrow \frac{C_{k,n+1} - C_{k,n}}{\Delta t} \Delta x \Delta y \Delta z \tag{E7.3.1}$$

The flux terms are discretized at the cell interface. Because transport occurs in the direction of flow, we will use upwind differences. In addition, we will use an explicit solution technique by discretizing our flux terms at the n (previous) time step:

$$uC\big|_{k+1/2,n} \Delta x \Delta y = [\![UC_{k,n}, -UC_{k+1,n}]\!] \Delta x \Delta y \tag{E7.3.2}$$

and

$$uC\big|_{k-1/2,n} \Delta x \Delta y = [\![UC_{k-1,n}, -UC_{k,n}]\!] \Delta x \Delta y \tag{E7.3.2}$$

Then,

$$\text{Flux in} - \text{Flux out} = ([\![UC_{k-1,n}, -UC_{k,n}]\!] - [\![UC_{k,n}, -UC_{k+1,n}]\!]) \Delta x \Delta y \tag{E7.3.3}$$

Now, rewriting equation (2.1) using our discretization and dividing by $\Delta x \Delta y \Delta z$:

$$\frac{C_{k,n+1} - C_{k,n}}{\Delta t} = [\![UC_{k-1,n}, -UC_{k,n}]\!] - [\![UC_{k,n}, -UC_{k+1,n}]\!] \tag{E7.3.3}$$

which can be rearranged to give an explicit equation for $C_{k,n+1}$:

$$C_{k,n+1} = Cou \left[\!\!\left[\frac{U}{|U|}, 0 \right]\!\!\right] C_{k-1,n} + (1 - Cou)\, C_{k,n} + Cou \left[\!\!\left[-\frac{U}{|U|}, 0 \right]\!\!\right] C_{k+1,n} \tag{E7.3.4}$$

where $Cou = |U|\,\Delta t/\Delta z$. A Courant number of 1 would result in a numerical solution that is an exact solution of the concentration front. A Courant number greater than 1 would result in instability and divergence from the solution. A Courant number less than 1 results in a front with numerical diffusion. We will use a Δz of 10 mm and a variable Δt to give solutions with different Courant numbers.

The breakthrough curve for different values of the Courant number is given in Figure E7.3.1. A lower Courant number, less than 1, adds more numerical diffusion to the solution. If the Courant number is greater than 1, the solution is unstable. This $Cou > 1$ "solution" is not shown in Figure E7.3.1 because it dwarfs the actual solution. Thus, for a purely convective problem, the Courant number needs to be close to 1, but not greater than 1, for an accurate solution. In addition to the value of the Courant number, the amount of numerical diffusion depends on the value of the term $U\Delta z$, which is the topic discussed next.

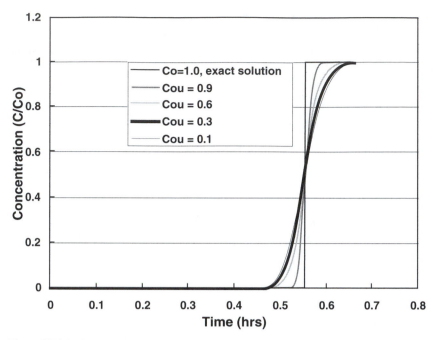

Figure E7.3.1. Computational solutions to the purely convective filter transport problem with varying Courant number.

Let us return to the discussion of computational transport routines, where each computational cell is the equivalent of a complete mix reactor. If we are putting together a computational mass transport routine, we could simply specify the size of the cells to match the diffusion/dispersion in the system. The number of well-mixed cells in an estuary or river, for example, could be calculated from equation (6.44), assuming a small Courant number. Then, the equivalent longitudinal dispersion coefficient for the system would be calculated from equation (6.44), as well, for a small Δt (Δt was infinitely small in equation 6.44):

$$D_L = \frac{UL}{2n} = \frac{UL}{2}\sigma^2 \quad \text{(small } Cou\text{)} \tag{7.11}$$

or, assigning $\Delta x = L/n$,

$$\Delta x = 2D_L/U \quad \text{(small } Cou\text{)}$$

Then, we can compute a "numerical dispersion coefficient," D_{num}:

$$D_{num} = \frac{1}{2} U \Delta x \quad \text{(small } Cou\text{)} \tag{7.12}$$

We could operate our computational transport model with only numerical dispersion (i.e., as a tanks-in-series model). It is often inconvenient to do so in environmental transport applications, however, because the cross-sectional mean velocity, U, can vary considerably over time and distance. Because the numerical dispersion coefficient depends on U, this would require that the number of cells be dependent on U. The number of cells, or Δx, is typically set such that D_{num} will be less than the true D_L.

EXAMPLE 7.4: *Comparison of solution to a front and a solution with dispersion*

The solution of Example 7.3 will be compared with an analytical solution of a diffusive front moving at velocity U, with $D = 1/2\, U\, \Delta z$. First, we must derive the analytical solution. This problem is similar to Example 2.10, with these exceptions: (1) convection must be added through a moving coordinate system, similar to that described in developing equation (2.36), and (2) a diffusion gradient will develop in both the $+z$- and $-z$-directions.

Then, the diffusion equation for the filter is

$$\frac{\partial C}{\partial t} + U\frac{\partial C}{\partial z} = D\frac{\partial^2 C}{\partial z^2} \tag{E7.4.1}$$

Using the moving coordinate system,

$$z^* = z - Ut \tag{E7.4.2}$$

Equation (E7.4.1) becomes

$$\frac{\partial C}{\partial t} = D\frac{\partial^2 C}{\partial z^{*2}} \tag{E7.4.3}$$

with boundary conditions:

1. $z^* = 0;\ C = C_0/2$
2. $t = 0, z > 0;\ C = 0$
3. $t = 0, z < 0:\ C = C_0$

Boundary condition 1 is chosen because of similarity in diffusion gradients in the $+z$- and $-z$-directions. Similar to Example 2.10, we will assign

$$\eta = \frac{z^*}{\sqrt{4Dt}} \tag{E7.4.4}$$

and equation (E7.4.3) becomes

$$\frac{d^2C}{d\eta^2} + 2\eta\frac{dC}{d\eta} = 0 \tag{E2.10.2}$$

with solution

$$C = \beta_1\, \text{erf}\,(\eta) + \beta_2 \tag{E7.4.5}$$

Now, at $\eta = 0$, $C = C_0/2$, thus

$$\beta_2 = C_0/2$$

At $\eta = \infty$, $C = 0$

$$0 = \beta_1 1 + C_0/2$$

or

$$\beta_1 = -C_0/2$$

Then, our solution to equation (E7.4.5) is

$$C = \frac{C_0}{2}\left[1 - \text{erf}\left(\frac{z - Ut}{\sqrt{4Dt}}\right)\right] = \frac{C_0}{2}\,\text{erfc}\left(\frac{z - Ut}{\sqrt{4Dt}}\right) \qquad \text{(E7.4.6)}$$

If we assign $D = U\Delta z/2$, equation (E7.4.6) becomes

$$\frac{C}{C_0} = \frac{1}{2}\,\text{erfc}\left(\frac{z - Ut}{\sqrt{2U\Delta zt}}\right) \qquad \text{(E7.4.7)}$$

The analytical solution of Equation (E7.4.7) is compared with the computational solution with a grid Courant number of 0.1 (and no diffusion coefficient) in Figure E7.4.1. It is seen that there is not much difference between the two.

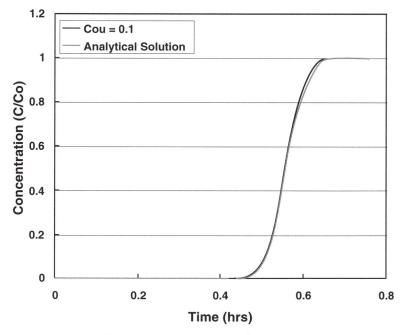

Figure E7.4.1. Solution to convective transport problem at $C_0 = 0.1$ and comparison with equation (E7.4.7) with $D = U\Delta z/2$. *Cou*, Courant number.

D. Explicit, Exponential Difference Solutions

Central differences are applied to diffusion problems, and upwind differences are applied to convective problems, but most cases have both diffusion and convection. This conundrum led Spaulding (1972) to develop exponential differences, which combines both central and upwind differences in an analytical solution of steady, one-dimensional convection and diffusion. Consider a control volume of length Δx, in a flow field of velocity U, and transporting a compound, C, at steady state with a diffusion coefficient, D. Then, the governing equation inside of the control volume is a simplification of Equation (2.14):

$$\frac{d}{dx}(UC) = \frac{d}{dx}\left(D\frac{dC}{dx}\right) \tag{7.13}$$

where we have used ordinary derivatives because x is the only independent variable. Assign $x = 0$ at the upstream side of the control volume. Then, the boundary conditions are:

1. At $x = 0$, $C = C_0$
2. At $x = \Delta x$, $C = C_1$

Integrating equation (7.15) once gives

$$D\frac{dC}{dx} = UC + \beta_1 \tag{7.14}$$

where β_1 is a constant of integration. Equation (7.14) can be written as a first-order linear differential equation:

$$\left(\lambda - \frac{U}{D}\right)C + \beta_1 = 0 \tag{7.15}$$

where λ is the d/dx operator. The solution to equation (7.15) is

$$C = C_c + C_p \tag{7.16}$$

where the complimentary solution is

$$C_c = \beta_2 e^{Ux/D} \tag{7.17}$$

The particular solution is found by applying boundary condition 1 to equation (7.16):

$$C_p = C_0 - \beta_2 \tag{7.18}$$

Then, applying boundary condition 2 to equation (7.16), we find β_2:

$$\beta_2 = \frac{C_1 - C_0}{1 - e^{U\Delta x/D}} \tag{7.19}$$

and equation (7.16) can now be expressed as

$$\frac{C - C_0}{C_1 - C_0} = \frac{1 - e^{Pe(x/\Delta x)}}{1 - e^{Pe}} \tag{7.20}$$

where $Pe = U\Delta x/D$ is the grid Peclet number. Equation (7.20) will be used to develop the exponential difference scheme, which transitions smoothly between upwind and central differences, and provides an exact solution for steady, one-dimensional convective-diffusive flow.

Now, the flux at the $i - 1/2$ interface is

$$J_{i-1/2} = UC_{i-1/2} - D \left.\frac{\partial C}{\partial x}\right|_{i-1/2} \qquad (7.21)$$

Inserting equation (7.20) into equation (7.21) gives

$$J_{i-1/2} = \frac{U}{\Delta x}\left(C_{i-1} + \frac{C_i - C_{i-1}}{1 - e^{Pe}}\right) \qquad (7.22)$$

and for the $i + 1/2$ interface:

$$J_{i+1/2} = \frac{U}{\Delta x}\left(C_i + \frac{C_{i+1} - C_i}{1 - e^{Pe}}\right) \qquad (7.23)$$

Then, equation (7.10) becomes

$$\frac{C_{i,n+1} - C_{i,n}}{\Delta t} = \frac{U}{\Delta x}\left(C_{i-1,n} + \frac{C_{i,n} - C_{i-1,n}}{1 - e^{Pe}}\right) - \frac{U}{\Delta x}\left(C_{i,n} + \frac{C_{i+1,n} - C_{i,n}}{1 - e^{Pe}}\right) + S_{i,n}$$

$$(7.24)$$

where subscript n indicates time step and subscript i indicates distance step in the x-direction. Equation (7.24) can be rearranged to an explicit equation for $C_{i,n+1}$:

$$C_{i,n+1} = \left(1 + Cou\frac{1 + e^{Pe}}{1 - e^{Pe}}\right)C_{i,n} - \frac{Cou}{1 - e^{Pe}}\left(\llbracket e^{Pe}, 1\rrbracket C_{i-1,n} + \llbracket e^{-Pe}, 1\rrbracket C_{i+1,n}\right) + S_{i,n}\Delta t$$

$$(7.25)$$

where a velocity in the $+x$-direction gives a positive Pe and one in the $-x$-direction gives a negative Pe. As before, $|U|\,\Delta t/\Delta x$ is always positive. Again, the term in brackets before $C_{i,n}$ should not be negative. That means that the Courant number criteria is

$$Cou \leq \frac{e^{Pe} - 1}{e^{Pe} + 1} \qquad (7.26)$$

In addition, we will still need to follow the stability criteria, $Di < 0.5$.

E. Implicit, Upwind Difference Solutions

The prior discretization of equation (2.1) uses control volumes with explicit differences. They are explicit because *only* the *accumulation* term contains a concentration at the $n + 1$ time step, resulting in an explicit equation for $C_{k,n+1}$ (equations (E7.1.4), (E7.2.5), (E7.3.4), and (7.25)). Another common option would be fully implicit (Laasonen) discretization where *flux rate* terms in equations (7.24) and (7.23) are computed at the $n + 1$ time increment, instead of the n increment. Fully implicit is generally preferred over Crank–Nicolson implicit ($UC_i = U\left(C_{i,n} + C_{i,n+1}\right)/2$)

because it is unconditionally stable, which is not the case for Crank–Nicolson implicit discretization. This results in an equation for $C_{k,n+1}$ that is more difficult to solve than the explicit differences, but allows for larger time increments (Δt). Then, equation (E7.3.4) becomes

$$(1 + Cou)C_{k,n+1} = Cou \left[\!\left[\frac{U}{|U|}, 0 \right]\!\right] C_{k-1,n+1} + C_{k,n} + Cou \left[\!\left[\frac{-U}{|U|}, 0 \right]\!\right] C_{k+1,n+1}$$

(7.27)

Because velocity will either be upwind or downwind, one of the two bracket terms will be zero, and equation (7.27) is solved as

$$C_{k,n+1} = \frac{Cou}{(1 + Cou)} \left[\!\left[\frac{U}{|U|}, 0 \right]\!\right] C_{k-1,n+1} + C_{k,n} + \frac{Cou}{(1 + Cou)} \left[\!\left[\frac{-U}{|U|}, 0 \right]\!\right] C_{k+1,n+1}$$

(7.28)

An example of the capabilities of implicit differences is provided in Figure 7.2, where the explicit difference computations with a grid Courant number of 0.1 are compared with the equivalent implicit difference computations with a grid Courant number of 10. There is more numerical diffusion at a grid Courant number of 10, but the explicit routine does not function at a grid Courant number greater than 1. At a grid Courant number of 0.1, the explicit and implicit techniques gave identical solutions. If one is willing to put up with some additional numerical diffusion, substantial computer time can often be saved with implicit differences. This time savings sometimes may not apply when nonlinear problems are to be solved in more than one dimension, because the solution of the implicit routine requires iterations that can reduce the difference in computational time.

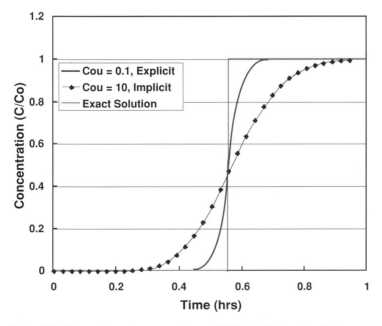

Figure 7.2. Comparison of computations with fully implicit and explicit upwind differences.

F. Implicit, Exponential Difference Solutions

With an implicit scheme applied to exponential differences, equation (7.24) becomes

$$\frac{C_{i,n+1} - C_{i,n}}{\Delta t} = \frac{U}{\Delta x}\left(C_{i-1,n+1} + \frac{C_{i,n+1} - C_{i-1,n+1}}{1 - e^{Pe}}\right)$$

$$- \frac{U}{\Delta x}\left(C_{i,n+1} + \frac{C_{i+1,n+1} - C_{i,n+1}}{1 - e^{Pe}}\right) + S_{i,n+1} \qquad (7.29)$$

Let's assume that we have zero-order and first-order source/sink terms (the most common case). Then,

$$S_{i,n+1} = k_0 + k_1 C_{i,n+1} \qquad (7.30)$$

Equation (7.30) is solved with a tridiagonal matrix algorithm as described in Patankar (1980). First reform equation (7.29) into

$$a C_{i-1,n+1} + b C_{i,n+1} + c C_{i+1,n+1} + z_{i,n} = 0 \qquad (7.31)$$

where

$$a = -Cou\left(\frac{[\![e^{Pe}, 1]\!]}{e^{Pe} - 1}\right)$$

$$b = 1 + Cou\left(\frac{e^{Pe} + 1}{e^{Pe} - 1}\right) - k_1 \Delta t$$

$$c = -Cou\left(\frac{[\![e^{-Pe}, 1]\!]}{e^{Pe} - 1}\right)$$

and

$$z_{i,n} = -k_0 \Delta t - C_{i,n}$$

Then, to solve for the concentration profile at $t = n + 1$, equation (7.31) is rewritten as

$$C_{i,n+1} + W_i C_{i+1,n+1} = G_i \qquad (7.32)$$

where, at all locations except the boundaries,

$$W_i = -\frac{c}{b + a W_{i-1}} \qquad (7.33)$$

$$G_i = \frac{a G_{i-1} + z_{i,n}}{b + a W_{i-1}} \qquad (7.34)$$

At $i = 1$, $C_{1,n+1}$ is known, and

$$W_2 = -\frac{c}{b} \qquad (7.35)$$

$$G_2 = \frac{a C_{1,n+1} + z_{1,n}}{b} \qquad (7.36)$$

At the far end of the domain, $i = I$, a constant gradient is normally a good assumption, where

$$C_{I+1,n+1} = 2C_{I,n+1} - C_{I-1,n+1} \tag{7.37}$$

Then,

$$W_I = 0$$

$$G_I = \frac{(a - c)G_{I-1} + z_{I,n}}{b - 2c + (a - c)W_{I-1}}$$

and finally,

$$C_{I,n+1} = G_I \tag{7.38}$$

The concentration profile at $t = n + 1$ is then found through back-substitution from $I = I - 1$ to $I = 2$. Rearranging equation (7.32):

$$C_{i,n+1} = G_i - W_i C_{i+1,n+1} \tag{7.39}$$

The explicit and implicit exponential difference techniques will be compared in Example 7.5.

EXAMPLE 7.5: *Comparison of explicit and implicit exponential differences with the exact solution*

The problem of Example 7.3 will again be solved with explicit and implicit exponential differences, and compared with the analytical solution, equation (E7.4.7). This solution is given in Figure E7.5.1. Note that the explicit solution is close to the analytical solution, but at a Courant number of 0.5, whereas the implicit solution could solve the problem with less accuracy at a Courant number of 5. In addition, the diffusion number of the explicit solution was 0.4, below the limit of $Di < 0.5$. The implicit solution does not need to meet this criteria and had $Di = 4$.

G. Problems

1. Solve Example 2.2 with a computational mass transport routine under the following conditions:

   Initial mass $= 5 \text{ kg/m}^2$
   Diffusion coefficient $= 10^{-9} \text{m}^2/\text{s}$

   Compare the result to the analytical solution for Example 2.2.

2. Assume that the solubility of a spilled compound for problem 2 in water is 5 kg/m^3, so that an impulse solution will not be accurate. The density of the spilled compound is slightly less than water, so it will float on the groundwater interface. How should these boundary conditions be handled in a computational routine?

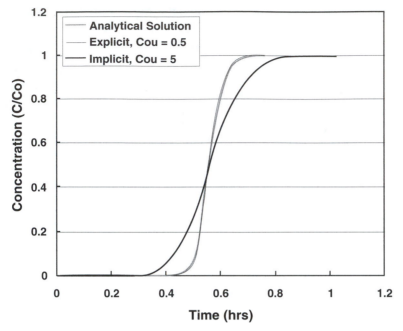

Figure E7.5.1. Comparison of explicit, implicit, and analytical solutions for the filter problem. $Pe = 1.25$ for explicit and implicit solutions. *Cou*, Courant number.

3. Solve problem 5, Chapter 2, using a central difference routine. How does your solution compare with an analytical solution? Make sure that your numerical solution is accurate at the lake interface and determine the flux over time of methylchloride into the lake.

4. Solve problem 5, Chapter 2, using an appropriate computational routine and compare the results to the analytical solution.

5. On leaving their military bases in Czechoslovakia, the military of the old Soviet Union dumped all chemicals out onto the ground in an unbelievable show of disrespect. At one location, 50,000 kg of solvent was dumped out of a storage tank onto saturated soil. The solvent formed a pool on the surface that was visible for more than 2 weeks. As part of a hazard assessment, the Czech engineers need to know the total mass of benzene (one compound composing the solvent) that was evaporated into the air versus time for the 2-week period. Estimate and plot this mass. The mean depth of the pool just after the spill was 10 cm. Incorporate the diffusion of benzene into the saturated soil. Assume that there is no turbulence in the spilled pool.

Data:

  Solvent is 20% benzene by weight
  Specific gravity of the solvent is 0.9
  Diffusivity of benzene in the solvent is 4×10^{-9} m^2/s

K_G for evaporation is large
Assume no sources or sinks
Diffusivity (effective) in soil $= 5 \times 10^{-10}$ m^2/s
$K_{ow} = 135$
$\rho_b/\varepsilon = 5$
$f = 0.01$

Test your computational mass transport routine before tackling this problem, with a known solution of some problem that has similar characteristics in time and space. Show this test when you write up your answer to the problem.

6. Solve problem 5, Chapter 5, with an appropriate numerical routine. Compare your solution to the analytical solution.

8   Interfacial Mass Transfer

A. Background

Interfacial transfer of chemicals provides an interesting twist to our chemical fate and transport investigations. Even though the flow is generally turbulent in both phases, there is no turbulence across the interface in the diffusive sublayer, and the problem becomes one of the rate of diffusion. In addition, temporal mean turbulence quantities, such as eddy diffusion coefficient, are less helpful to us now. The unsteady character of turbulence near the diffusive sublayer is crucial to understanding and characterizing interfacial transport processes.

1. Two-Phase Mass Balance

We will discuss a closed container containing air and water, illustrated in Figure 8.1. The water in the container contains a given amount of the phenol, hydroxylbenzene, or C_6H_5OH.

The mass flux rate is given as

$$\frac{dm_{hb}}{dt} = \Psi_a \frac{d\,C_{hb,a}}{dt} = \Psi_w \frac{dC_{hb,w}}{dt} = J_{hb} \tag{8.1}$$

where m is mass, Ψ is volume, J is a temporal mean flux rate/unit surface area, the subscript hb indicates hydroxylbenzene, and subscripts a and w indicate air and water, respectively. Flux of chemical occurs because there is a *concentration gradient* at the interface, as illustrated in Figure 8.2.

The flux rate will be constant with z, unless we are very close to equilibrium or there are sources or sinks of hydroxylbenzene in the container. At the interface, equilibrium is assumed between the air and water phases. This equilibrium may only exist for a thickness of a few molecules, but is assumed to occur quickly compared with the time scale of interest. The concentrations at this air–water interface are related by the following equation:

$$\frac{C_{hb,a}^*}{C_{hb,w}^*} = H_{hb} \tag{8.2}$$

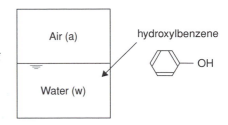

Figure 8.1. Closed container used to study the mass balance of hydroxylbenzene.

where H_{hb} is the hydroxylbenzene equilibrium partitioning coefficient for air–water equilibrium, also called Henry's law constant, and the asterisks indicate the concentrations at equilibrium. Now, because J_{hb} is constant over vertical distance,

$$J_{hb} = \text{Const.} = -(D_{hba} + \varepsilon_{za})\frac{\partial C_{hba}}{\partial z} = -(D_{hbw} + \varepsilon_{zw})\frac{\partial C_{hbw}}{\partial z} \qquad (8.3)$$

where D is the diffusion coefficient and ε is the eddy diffusion coefficient.

Now, let us examine equation (8.3) with the knowledge that J_{hb} is constant over space. Away from the interface, we know that $\partial C_{hba}/\partial z$ and $\partial C_{hbw}/\partial z$ are both small. This means that, for J_{hb} to be constant, $D_{hba} + \varepsilon_z$ and $D_{hbw} + \varepsilon_z$ must be large. The inverse is also true. When $\partial C_{hba}/\partial z$ and $\partial C_{hbw}/\partial z$ are large, $D_{hba} + \varepsilon_z$ and $D_{hbw} + \varepsilon_z$ must be small. The only thing that can vary with distance that dramatically in the $D + \varepsilon_z$ terms is ε_z. Thus, the ε_z versus z relationship is seen in the concentration profile.

We will now adjust our position to within a few microns of the interface and examine the flux. *Flux is molecular at the interface* between two phases, as illustrated in Figure 8.3. This means that ε_z is zero and the only active process is diffusion. Note the difference in the order of magnitudes of D and ε given in Figure 8.3, and recall the resistance concepts that we discussed in Chapter 2. The resistance to mass transfer is much larger near the interface than away from the interface, so the controlling process would tend to be transfer in the diffusive sublayer.

In the diffusive sublayer, flux is due purely to diffusion. Then, we can utilize the fact that mean flux is constant with distance:

$$\overline{J_{hb}} = -D_{hb,a}\frac{\overline{\partial C_{hba}}}{\partial z} = -D_{hb,w}\frac{\overline{\partial C_{hbw}}}{\partial z} \qquad (8.4)$$

Figure 8.2. Illustration of flux rate across a small portion of the water surface in Figure 8.1.

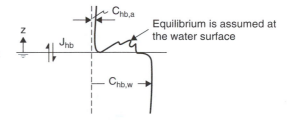

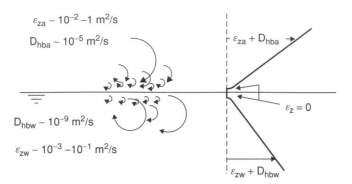

Figure 8.3. Illustration of diffusive flux near the air–water interface.

Because $\overline{J_{hb}}$ is constant and D_{hba} and D_{hbw} are also constant, assuming the thermo-dynamic parameters such as temperature and pressure do not change, the solution to equation (8.4) is

$$\overline{C_{hb,a}} = \beta_0 + \beta_1 z \tag{8.5}$$

and

$$\overline{C_{hb,w}} = \beta_2 + \beta_3 z \tag{8.6}$$

where $z = 0$ at the interface and the interface concentrations are β_0 and β_2. We can put the concentration of hydroxylbenzene in the air and water on the same scale using the Henry's law constant (equilibrium partitioning) by dividing the air concentration by H_{hb}, as illustrated in Figure 8.4.

Then, equation (8.4) may be written as

$$J_{hb} = -D_{hb,a} H_{hb} \frac{\partial(C_{hb,a}/H_{hb})}{\partial z} = -D_{hb,w} \frac{\partial C_{hb,w}}{\partial z} \tag{8.7}$$

that does not have a discontinuity in concentration (C_{hba}/H_{hb} or C_{hbw}) at the interface. From equation (8.7), we can see that compounds with a large value of H will have a small concentration gradient, $\partial(C/H)/\partial z$, and the equivalent diffusion coefficient, $D_a H$, will be large, compared with D_w. These compounds include oxygen, nitrogen,

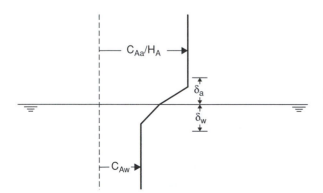

Figure 8.4. The result of putting concentrations on the same scale using Henry's law constant.

methane, carbon tetrachloride, tetra chloroethylene, lower molecular weight polychlorinated biphenols, some aromatic hydrocarbons, C_4H_6, and so on, and are classified as *volatile compounds*. Compounds with a small value of H will have a large
value of concentration gradient, $\partial(C/H)/\partial z$, and the equivalent diffusion coefficient,
$D_a H$, will be small, relative to D_w. These compounds include compounds that have
an attraction to dissolution in water, such as acrylic acid, benzidiene, benzoic acid,
and water itself. They are classified as *nonvolatile compounds*. Finally, there are those
compounds in which $D_a H$ is of the same order of magnitude as D_w, such as phenan-
threne, naphthaline, heptanone, and the pesticide aldrin. These are called *semivolatile
compounds*.

2. Bulk Transfer Rate Coefficient

As noted previously, most environmental flows are turbulent. The diffusive sublayer,
where only diffusion acts to transport mass and the concentration profile is linear,
is typically between $10\,\mu m$ and $1\,mm$ thick. Measurements within this sublayer are
not usually feasible. Thus, the interfacial flux is typically expresses as a *bulk transfer
coefficient*.

$$J_A = K_B\,(C_{Aa}^{\infty} - C_{Aw}^{\infty}/H_A) \tag{8.8}$$

where K_B is the bulk transfer coefficient, C_{Aa}^{∞} is the concentration of compound A
away from the diffusive sublayer in the air phase, and C_{Aw}^{∞} is the concentration of
compound A away from the diffusive sublayer in the water phase.

The *resistance* to interfacial transfer can be visualized as $1/K_B$:

$$\frac{1}{K_B} = \frac{1}{K_L} + \frac{1}{H_A K_G} \tag{8.9}$$

where K_L is the liquid film coefficient and K_G is the gas film coefficient. Equation (8.9) indicates that K_B is dependent on the chemical being transferred, as it
contains H_A. The coefficients K_L and K_G are less dependent on the properties of the
gas. Henry's law constant for various compounds of environmental interest and the
percent resistance to transfer in the liquid phase for $K_G/K_L = 20$, 50, and 150 are
given in Appendix A–6. A K_G/K_L ratio of 20 to 50 corresponds to bubble-influenced
transfer (Munz and Roberts, 1984), whereas a ratio of 150 is more appropriate for
free-surface transfer (Mackay and Yuen, 1983).

Equations (8.4), (8.8), and (8.9) can be combined with the assumption that all
resistance to transfer occurs in the diffusive sublayer, as illustrated in Figure 8.5, to
show that

$$K_L = \frac{D_{Aw}}{\delta_w} \tag{8.10}$$

and

$$K_G = \frac{D_{Aa}}{\delta_a} \tag{8.11}$$

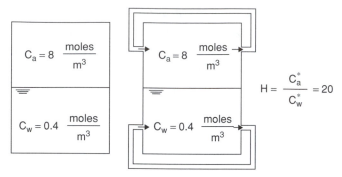

Figure 8.5. Equilibrium = steady state in both the stagnant system and that with flow.

It may seem as though we have abandoned our statement that the unsteady aspects of the interaction of the diffusive sublayer and turbulence are paramount, because K_B, K_L, and K_G are all bulk quantities. However, the unsteady relationships that exist will still be brought into the analysis of equation (8.10) and (8.11) (i.e., $\delta = \delta(D, \text{turbulance})$). This relatively simple characterization provides for most of the research regarding interfacial transport rates.

B. Equilibrium Partitioning

Partitioning coefficients describe the phase equilibrium of one solute between the two phases. At equilibrium, there is no net transfer. When there is a net transfer, it is proportional to the difference from equilibrium. Equilibrium is therefore always important, because it is the result of transport; it is where the chemical concentrations are going. Example 8.1 will show us that even a common term, such as relative humidity, can be related to Henry's law.

EXAMPLE 8.1: *Air–water equilibrium of H_2O*

On a humid day in August, the radio announcer says that the relative humidity is 70%. Relate this to Henry's law constant for water. At the same temperature, why does 70% relative humidity feel so much warmer than 30% relative humidity?

As illustrated in Figure E8.1.1, a relative humidity of 100% is at air–water equilibrium for H_2O. The relative humidity does not usually rise beyond 100% because water vapor condenses. At a relative humidity below 100%, water evaporates. At 100% relative humidity, the partial pressure of water vapor in the air is equal to the pressure at equilibrium, also called the vapor pressure.

The rate of vaporization (evaporation) is given by

$$J_{H_2O} = K_{BH_2O}\left(C_{H_2O,w} - C_{H_2O,a}/H_{H_2O}\right) \tag{E8.1.1}$$

Note that

$$C_{H_2O,w} = \rho_{H_2O}$$

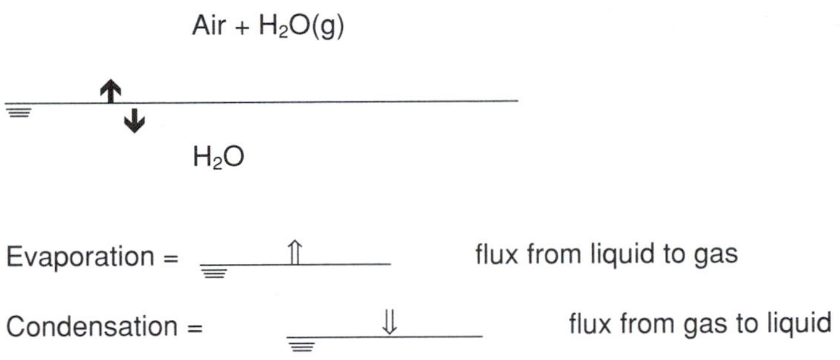

Air + H₂O(g)

H₂O

Evaporation = ⟰ flux from liquid to gas

Condensation = ⟱ flux from gas to liquid

Figure E8.1.1. Illustration of the air–water equilibrium of H_2O.

Because there is no net transfer of water at equilibrium,

$$H_{H_2O} = \frac{C_{H_2O,a}}{\rho_{H_2O}} = \frac{P_v}{P_{air}} \frac{M_{H_2O}}{M_{air}} \frac{\rho_{air}}{\rho_{H_2O}} \tag{E8.1.2}$$

where P_v is vapor pressure and P_{air} is the pressure of the air. Plugging in some typical numbers ($P_v/P_{air} \sim 0.02$, $M_{H_2O} = 18\,g/mole$, $M_{air} = 29\,g/mole$, $\rho_{air} \sim 1.2\,kg/m^3$, and $\rho_{H_2O} = 1{,}000\,kg/m^3$) results in a value of H_{H_2O} on the order of 10^{-5}. Applying equation (8.9) to water vapor gives

$$\frac{1}{K_{B,H_2O}} = \frac{1}{K_{L,H_2O}} + \frac{1}{H_{H_2O}\,K_{G,H_2O}} \tag{8.9}$$

Because the Henry's law constant for water is on the order of 10^{-5}, $K_{B,H_2O} \Rightarrow H_{H_2O} K_{G,H_2O}$, and equation (E8.1.1) becomes

$$J_{H_2O} = K_{G,H_2O}\left(H_{H_2O}C_{H_2O,w} - C_{H_2O,a}\right) = H_{H_2O}C_{H_2O,w}K_{GH_2O}(1 - RH/100) \tag{E8.1.3}$$

where RH is relative humidity. Evaporation is a large source of cooling, because the heat of vaporization for water is approximately 540 cal/g. Equation (E8.1.3) indicates that the rate of vaporization, H_{H_2O}, is directly related to relative humidity. Thus, the rate of cooling of our skin is also directly related to relative humidity, because we are animals who sweat. Equation (E8.1.3) indicates that evaporative water flux would be much larger at $RH = 30\%$ than at $RH = 70\%$. Animals who do not sweat (e.g., reptiles) are largely unaffected by relative humidity. Your car's cooling system would also be unaffected by relative humidity.

Relative humidity (RH) is related to Henry's law constant through the relationship implied in equation (E8.1.3), or

$$RH = 100C_{H_2Oa}/(H_{H_2O}\rho_{H_2O}) \tag{E8.1.4}$$

A relative humidity of 70% means that we are at 70% of air–water equilibrium or $C_a/\rho_w = 0.7\,H_{H_2O}$.

1. Air–Water Partitioning Coefficient

In 1803, William Henry observed that the quantity of a gas that will dissolve into water is directly proportional to its partial pressure. This observation resulted in Henry's law and Henry's law constant, H. For a given compound, indicated by A,

$$H_A = \frac{P_A}{C_{A,w}} \text{(Atm m}^3/\text{g)}$$ (8.12)

or

$$H_A = \frac{C_{Aa}}{C_{Aw}} (-)$$ (8.13)

or

$$H_A = \frac{P_A}{X_A} \quad \text{(Atm, where } X_A = \text{mole fraction)}$$ (8.14)

where the units of H_A are indicated in parentheses. Henry's law constant for the compound of interest can be found in Appendix A–6, in the *Handbook of Chemistry and Physics* (CRC, 2005), or estimated from solubility and vapor pressure estimates/measurements (Lyman et al., 1990). In general, if you are interested in determining H_A for a compound that is a gas at atmospheric pressure,

$$H_A = \frac{\text{Partial pressure of A}}{\text{Solubility of A in water}}$$ (8.15)

For compounds that are a liquid at atmospheric pressure,

$$H_A = \frac{P_{vA}}{C_{Aw}^*}$$ (8.16)

where P_{vA} is the vapor pressure of the pure liquid compound, and C_{Aw}^* is the solubility of the compound in water.

EXAMPLE 8.2: *Determination of Henry's law constants*

You need the Henry's law constants for benzene and DDT. Determine these values in the units of atm m^3/g, atm m^3/mole, and dimensionless units.

Both DDT and benzene are liquids at room temperature. The required data for these two compounds is given in Table E8.2.1.

Benzene:

$$H_b = \frac{P_v}{C_\infty^*} = \frac{12,700 \text{ N/m}^3}{(1.01 \times 10^5 \text{ N/m}^2 - \text{atm})(1,780 \text{ g/m}^3)} = 7.0 \times 10^{-5} \text{ atm m}^3/\text{g}$$ (E8.2.1)

where the unit conversion 1 atm $= 1.01 \times 10^5$ N/m^2 has been used.

$$H_b = 7.0 \times 10^{-5} \text{ atm m}^3/\text{g} (78 \text{ g/mole}) = 5.5 \times 10^{-3} \text{ atm m}^3/\text{mole}$$ (E8.2.2)

$$H_b(-) = \frac{H_b(\text{atm m}^3/\text{mole})}{RT} = \frac{5.5 \times 10^{-3} \text{ atm m}^3/\text{mole}}{8.21 \times 10^{-5} \frac{\text{atm m}^3}{^\circ\text{K mole}} (298^\circ\text{K})}$$ (E8.2.3)

Table E8.2.1: *Required data for Example 9.2*

Chemical	Mol. wt.	Solubility in water (25°C)	Vapor pressure (25°C)
Benzene (*b*)	78	1,780 g/m³	12,700 N/m²
DDT	354	0.003 g/m³	2×10^{-5} N/m²
Symbol	M	C_w^*	P_V

such that

$$H_b(-) = 0.22 \qquad (E8.2.4)$$

We can thus compute a universal conversion of Henry's constant units from equations (E8.2.2) and (E8.2.3):

$$H(\text{atm m}^3/\text{mole}) = H(-)/40.9 \qquad (E8.2.5)$$

DDT:

$$H_{DDT} = \frac{2 \times 10^{-5} \text{ N/m}^2}{1.01 \text{ N/m}^2 \text{ atm}(0.003 \text{ g/m}^3)} = 6.6 \times 10^{-8} \frac{\text{atm m}^3}{\text{g}} \qquad (E8.2.6)$$

$$H_{DDT} = 6.6 \times 10^{-8} \frac{\text{atm m}^3}{\text{g}} (354 \text{ g /mole}) = 2.34 \times 10^{-5} \frac{\text{atm m}^3}{\text{mole}} \qquad (E8.2.7)$$

$$H_{DDT}(-) = \frac{2.34 \times 10^{-5} \text{ atm m}^3/\text{g}}{8.21 \times 10^{-5} \frac{\text{atm m}^3}{°\text{K mole}}(298°\text{K})} = 9.6 \times 10^{-4} \qquad (E8.2.8)$$

and

$$H_{DDT}\left(\frac{\text{atm m}^3}{\text{g}}\right) = \frac{H_{DDT}(-)}{14,500} \qquad (E8.2.9)$$

Note that Henry's law constant for DDT is much less than for benzene.

Henry's law constant is highly dependent on temperature. For example, oxygen has a Henry's law constant of 37 atm m³/g at 20°C and 70 atm m³/g at 0.1°C. The temperature effect on H is typically modeled with a form of the van't Hoff equation (see Hornix and Mannaerts, 2001).

$$H = \exp(A - B/T) \qquad (8.17)$$

where T is °K. Some temperature regressions on coefficients A and B in equation (8.17) are given in Table 8.1. Techniques to predict the coefficients have been developed by Nirmalakhandan et al. (1997).

Table 8.1: *Results of regressions on H vs. T for the determination of A and B in equation (8.17)*

| Compound | $H = \exp(A - B/T)$ | |
	A	B
benzene	11.8	3964
carbon tetrachloride	15.0	4438
chlorobenzene	9.6	3466
1-chlorobutane	11.3	3482
2-chlorobutane	15.1	4499
chloroform	11.8	4046
o-chlorotoluene	10.0	3545
1,4-dichlorobutane	6.6	3128
1,1-dichloroethylene	15.9	4618
1,2-dichloropropane	12.4	4333
1,3-dichloropropane	9.9	3917
ethylenedichloride	8.9	3539
methlylene chloride	10.2	3645
tetrachloroethane	7.40	3477
s-tetrachloroethylene	15.5	4735
1,1,1-trichloroethane	14.5	4375
1,1,2-trichloroethane	9.0	3690
trichloroethane	14.7	4647
1,2,3-trichloropropane	7.4	3477
toluene	11.3	3751

H is in nondimensional concentration units (Leighton and Calo, 1981)

2. Equilibrium or Steady State?

Equilibrium is *not* steady state and vice-versa. Sometimes they occur together, and other times they do not. A few quick examples will be used to illustrate this point.

First, consider two closed systems, illustrated in Figure 8.5, and let us consider oxygen, with a mass-related dimensionless Henry's law constant of 20. Equilibrium will occur simultaneously with steady state in these two systems.

Now, consider an air–water system with a given inflow concentration and an outflow concentration that results from the mass transfer in the system and the inflow. If the system were to run for a long time, steady state would be achieved, although the system is nowhere near equilibrium, as shown in Figure 8.6.

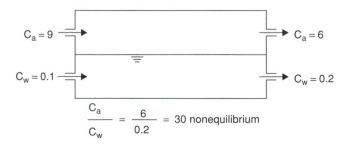

Figure 8.6. Illustration of a system at steady state but *not* at equilibrium.

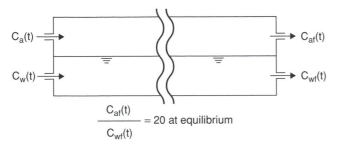

Figure 8.7. An infinitely long reactor with time-dependent inflow concentrations used to show the case when equilibrium is approached, but not steady state.

Finally, let us look at a system that approaches equilibrium, but is not at steady state. If we had an input concentration that is a function of time, and an infinitely long reactor, the system would never be at steady state because of the temporal variations in the inflow. The outflow, however, would always approach equilibrium, regardless of the inflow concentrations because of the reactor length. This is shown in Figure 8.7.

3. Partitioning to Other Phases in Contact with Water

Equilibrium partitioning to other phases (e.g., particulates, sediments, the biota, and oils) is generally predicted with the octanol-water partitioning coefficient, K_{ow}. We previously discussed octanol–water partitioning coefficients in Section 2.B, when we needed to derive a retardation coefficient. K_{ow} is important because it represents the tendency of a chemical to partition between an organic phase (fish, organic particles, organics in the sediment) and water. In general. a compound is considered hydrophilic if $K_{ow} < 10$, and hydrophobic if $K_{ow} > 10^4$. We will become accustomed to the use of K_{ow} through the next example.

EXAMPLE 8.3: *Experimental determination of the K_{ow} of aldrin*

Aldrin is a pesticide used mainly on crops of feed corn. We will first mix a few grams of aldrin into the octanol. Then, we will pour the octanol/aldrin mixture slowly into some water in a container and seal the container, as shown in Figure E8.3.1. While gently shaking the container, we will extract samples of the water over time to perform our analysis. The sample data will be fit to the equation for a first-order and zero-order process:

$$\left(\frac{C_A - C_A^*}{C_{A_0} - C_A^*}\right) = e^{-kt} \tag{E8.3.1}$$

where the subscript A indicates aldrin, k is some rate coefficient, C_{A_0} is the initial aldrin concentration in water, and C_A^* is the concentration at equilibrium. Equation (E8.3.1) may then be used to determine C_A^* through a least-squares fit on coefficients

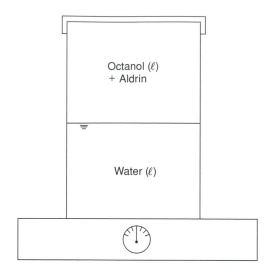

Figure E8.3.1. View of the octanol–water container on a shake table.

k and C_A^*. A similar operation could be performed with measurements taken from octanol samples, if the concentration in the octanol were to change significantly.

In terms of K_{ow}, the end result is

$$K_{ow} = \frac{C_{Aoct}^*}{C_{Aw}^*} = 13.5 \pm 10\text{--}20\% \tag{E8.3.2}$$

$$\log K_{ow} = 1.13 \pm 0.1\text{--}0.2 \tag{E8.3.4}$$

One convenient aspect of the octanol–water partitioning coefficient is that it is *not* sensitive to temperature. K_{ow} typically varies between 0.001 and 0.01 log units per °C. This means that the maximum difference for any compound, when comparing K_{ow} at 5°C to one at 25°C, is 0.2 log units, or a factor of 1.6.

Bioconcentration Factor in Aquatic Organisms. The bioconcentration factor (BCF) indicates the degree to which a chemical may accumulate in aquatic organisms, such as fish, clams, and zooplankton. It is given by the equation

$$\text{BCF} = \frac{\text{g chemical/g weight of organism}}{\text{g chemical/g of water}} \tag{8.18}$$

and because 1 g of water occupies approximately 1 cm³:

$$\text{BCF} = \frac{\text{g chemical/g weight of organism}}{\text{g chemical/cm}^3 \text{ of water}} \tag{8.19}$$

Lyman et al. (1990) recommend the equation of Veith et al. (1979):

$$\log(\text{BCF}) = 0.76 \log K_{ow} - 0.23 \tag{8.20}$$

Multimedia Partition Calculations. The environment, of course, contains all of the media that we have discussed: air, water, sediment, particles, living organisms, etc. The goal of multimedia partition calculations is to determine where the chemicals placed in the environment will eventually go. We will perform a multimedia calculation in Example 8.4. A sealed fish tank is chosen because it is a microcosm and does not require boundary conditions, such as air concentration.

EXAMPLE 8.4: *Multimedia partition calculations*

A large, sealed fish tank – consisting of $20\,m^3$ of water, $10\,m^3$ of air, $1\,m^3$ of sediment, and $0.2\,m^3$ of fish and other biological organisms – has 100 g of benzene accidentally spilled into it. What will be the eventual (equilibrium) concentrations and mass in each phase? The fraction of organic carbon in the sediments has been determined to be 0.01. The density of the sediments has been determined to be $2\,g/cm^3$. The octanol–water partitioning coefficient for benzene is 138.

We will use the subscripts b for benzene, a for air, w for water, s for sediments, and f for fish. We will also attempt to isolate one variable from the four media and have chosen the concentration in the water. Now, a mass balance on the 100 g of benzene gives

$$m_b = m_{ba} + m_{bw} + m_{bs} + m_{bf} = 100\,g \tag{E8.4.1}$$

where

$$m_{ba} = \frac{P_{ba}^* \rho_a V_a}{P} \tag{E8.4.2}$$

In equation (E8.4.2), P_{ba}^* is the partial pressure of benzene at equilibrium with the water, ρ_a is the density of the air ($1.2\,kg/m^3$), Va is the volume of the air ($10\,m^3$), and P is the air pressure in the tank ($\sim 1\,atm$). Before we can determine the partial pressure of benzene, we need to consider the water, where we will try to isolate our one variable, C_{bw}. In the water,

$$m_{bw} = C_{bw}^* V_w \tag{E8.4.3}$$

where C_{bw}^* is the equilibrium concentration in water and V_w is the volume of water ($20\,m^3$). For benzene, Henry's law constant is $6.9 \times 10^{-5} atm - m^3/g$. Then,

$$H_b = \frac{P_{ba}}{C_{bw}^*} \tag{E8.4.4}$$

and equation (E8.4.2) becomes

$$m_{ba} = C_{bw} H_b \rho_a V_a / P \tag{E8.4.5}$$

In the sediment, the equilibrium concentration of an organic compound adsorbed to the sediment is dependent on the fraction by weight of organic carbon.

$$m_{bs} = C^*_{bs} \rho_s V_s = C^*_{bw} K_d \rho_s V_s \tag{E8.4.6}$$

where ρ_s is the bulk density of the sediment ($2\,\mathrm{g/cm^3}$), V_s is the volume of the sediment ($1\,\mathrm{m^3}$), and C^*_{As} is the equilibrium concentration in the sediment. Using equation (2.34) (Karikhoff et al., 1979):

$$K_d = Bf K_{ow} = 0.41\ \mathrm{cm^3/g}\ (0.01)(134) \tag{E8.4.7}$$

or

$$K_d = 3.4\ \frac{\mathrm{g\ adsorbed/g\ sediment}}{\mathrm{g/cm^3}} \tag{E8.4.8}$$

The last item would be the fish and the other organisms in the tank. The mass of benzene in the fish at equilibrium is given by the equation

$$m_{bf} = \mathrm{BCF}_f \rho_f V_f C^*_{bw} \tag{E8.4.9}$$

where ρ_f is the density of the fish (assumed to be the same as water), V_f is the volume of the fish ($0.2\,\mathrm{m^3}$), and C^*_{bf} is the equilibrium concentration in the fish and other organisms. Applying equation (8.20) gives

$$\log \mathrm{BCF}_f = 0.76(2.13) - 0.23 = 1.39 \tag{E8.4.10}$$

or $\mathrm{BCF}_A = 24.5$.

Now, we substitute all of the equations for m_{bi} into equation (E8.4.1) to get

$$m_b = C^*_{bw}(H_b \rho_a \overline{V}_a + \overline{V}_b + k_d \rho_s \overline{V}_s + \mathrm{BCF}_f \rho_f \overline{V}_f) = 100\ \mathrm{g} \tag{E8.4.11}$$

where the only unknown is C^*_{bw}.

The previous equations given can be used to determine the mass and concentration of benzene in each media. These are given in Table E8.4.1. Often there is one or more concentrations that surprise the investigator.

Table E8.4.1: *Mass and concentration of benzene in each phase of the fish tank*

Media	Mass of benzene (g)	Concentration of benzene
Air	2.6	2.1×10^{-4} atm
Water	62	$3.1\,\mathrm{g/m^3}$
Sediment	21	$21\,\mathrm{g/m^3}$
Fish	15	$76\,\mathrm{g/m^3}$

C. Unsteady Diffusion Away from an Interface

The concentration boundary layer is rarely at steady state. The only transport mechanism away from the interface is diffusion, and diffusion is a slow process. We will practice dealing with unsteady diffusion away from an interface with Example 8.5.

EXAMPLE 8.5: *Unsteady diffusion away from an interface*

Water with 0.55 g/m^3 toxaphene is in contact with quiescent air. Estimate the growth of the concentration boundary layer in the air over time.

Toxaphene, or polychlorocamphene ($C_{10}H_{10}Cl_8$), is a pesticide for cotton. It has no significant biodegradation or photolysis reactions in the environment, so it is a persistent chemical. Henry's law constant for toxaphene is $H_t = 6.0 \times 10^{-6}$ atm m^3/mole. The initial conditions are given in Figure E8.5.1.

An analytical solution is known to this problem under two conditions:

1. the resistance to transport in one phase is negligible, compared with the resistance in the other phase, or
2. the diffusivities in both phases are identical.

Since condition 2 cannot be valid for an air–water transfer, we will look into the possibility that condition 1 may be assumed.

To determine if either side of the interface controls transport of toxaphene across the interface, we need to look at the ratio of transfer coefficients on either side. Mackay and Yuen (1983) suggest that K_G/K_L values of between 50 and 300 are reasonable. We will use $K_G/K_L = 150$. Then, the resistance to transfer is given as

$$\frac{1}{K_B} = \frac{RT}{H_t K_G} + \frac{1}{K_L} = \frac{RT}{H_t(150)K_L} + \frac{1}{K_L} \qquad \text{(E8.5.1)}$$

where R is the universal constant for the ideal gas law and T is absolute temperature. These parameters convert Henry's law constant from atm-m^3/mole to

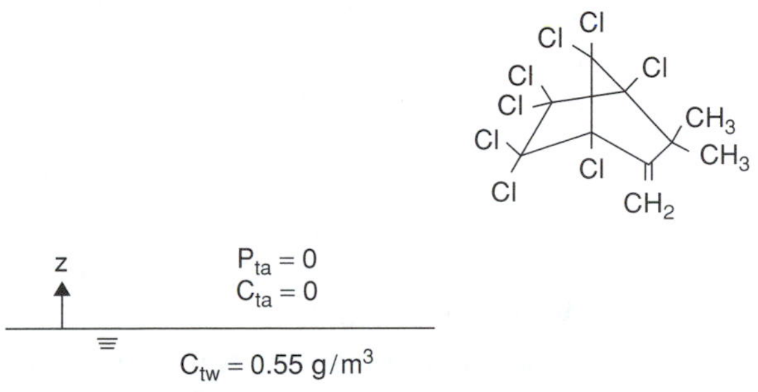

Figure E8.5.1. Initial conditions for the unsteady toxaphene flux problem.

z

$P_{ta} = 0$
$C_{ta} = 0$

$C_{tw} = 0.55$ g/m^3

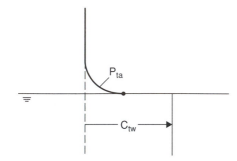

Figure E8.5.2. Concentration profiles for the example problem, assuming that $K_L \gg H_t H_G$.

dimensionless units. Substituting the numbers for toxaphene into equation (E8.5.1) results in

$$\frac{1}{K_B} = \frac{1}{K_L}\left[\frac{8.2 \times 10^{-5}\dfrac{\text{atm m}^3}{\text{mole }^\circ\text{K}}(293^\circ\text{K})}{6 \times 10^{-6}\dfrac{\text{atm m}^3}{\text{mole}}(150)} + 1\right] \tag{E8.5.2}$$

or

$$\frac{1}{K_B} = \frac{1}{K_L}(\underset{\substack{\uparrow \\ \textbf{air} \\ \textbf{resistance}}}{27} + \underset{\substack{\uparrow \\ \textbf{water} \\ \textbf{resistance}}}{1}) \cong \frac{RT}{H_t K_G} \tag{E8.5.3}$$

The air phase resistance is estimated to be 27 times the water phase resistance in equation (E8.5.3). Thus, the results that we get assuming the water phase resistance to be zero are close to those that we would get assuming that they will be 1/27 the total resistance. This indicates that toxaphene is a gas-phase (nonvolatile) controlled compound, as long as K_G/K_L is equal to 150. A diagram of the assumed concentration profiles is given in Figure E8.5.2.

In the air phase, the governing equation is

$$\frac{\partial P_t}{\partial t} = D_{ta}\frac{\partial^2 P_t}{\partial z^2} \tag{E8.5.4}$$

with boundary conditions:

1. At $t = 0$ and $z > 0$, $P_t = 0$
2. At $t > 0$ and $z = 0$, $P_t = P_t^*$

where P_t is the partial pressure of toxaphene in air and P_t^* is the equilibrium concentration. Equation (E8.5.4) and the boundary conditions are similar to Example 2.10, with a solution that is also similar:

$$P_t = \beta_1 + \beta_2 \,\text{erf}\left(\frac{z}{\sqrt{4D_{ta}t}}\right) \tag{E8.5.5}$$

where β_1 and β_2 are found from boundary conditions.

We will now apply the boundary conditions. Boundary condition 1 gives $\beta_1 = -\beta_2$. Boundary condition 2 gives $\beta_1 = P_t^*$ and $\beta_2 = -P_t^*$. Our toxaphene problem is now

Table E8.5.1: *Numerical values of P_t/P_t^* for toxaphene from equation E8.5.6*

Distance from interface	Time			
	10 s	100 s	10^3 s (15 min)	10^5 s (28 hrs)
100 μm	0.99	0.99	0.99	0.99
1 mm	0.92	0.98	0.99	0.99
1 cm	0.69	0.76	0.93	0.99
10 cm	0	0	0.32	0.92
1 m	0	0	0	0.32
δ_c	0.03 m	0.08 m	0.26 m	2.59 m

similar to Example 27. Applying the variables and parameters in this problem, the solution can be found in Example 2.7:

$$\frac{P_t}{P_t^*} = 1 - \mathrm{erf}\left(\frac{z}{\sqrt{4D_{ta}\,t}}\right) \tag{E8.5.6}$$

The Hayduk–Laudie relationship, described in Chapter 3, allows us to compute a value of $D_{ta} = 5 \times 10^{-6}\,\mathrm{m}^2/\mathrm{s}$. We can determine various values of P_t/P_t^*, as given in Table E8.5.1.

All that is left is to determine the equilibrium partial pressure for toluene, P_t^*:

$$P_t^* = \frac{C_{tw}^*(\mathrm{g/m}^3)\,H_t(\mathrm{atm-m}^3/\mathrm{mole})}{M_t(\mathrm{g/mole})} \tag{E8.5.7}$$

or

$$P_t^* = \frac{55\,\mathrm{g/m}^3(6 \times 10^{-6}\,\mathrm{atm\ m}^3/\mathrm{mole})}{414\,\mathrm{g/mole}} = 7.97 \times 10^{-7}\,\mathrm{atm} \tag{E8.5.8}$$

As long as the water concentration is 55g/m^3, the equilibrium partial pressure of toxaphene will be 7.97×10^{-7} atm.

The edge of the concentration boundary layer is typically determined to be where $P_t/P_t^* = 0.01$. That would be where erf $[z/(4\,D_{ta}t)^{1/2}] = 0.99$ or where $z = 3.66(D_{ta}t)^{1/2}$. The thickness of the concentration boundary layer, δ_c, is also given in Table E8.5.1 for each time. At 15 min and 28 hrs, of course, the value of δ_c is not realistic because the atmosphere is not quiescent (stagnant), and turbulence would mix the toxaphene into the surrounding air.

D. Interaction of the Diffusive Boundary Layer and Turbulence

In Example 8.5, we saw how the diffusive boundary layer could grow. The boundary never achieves 1 m in thickness or even 5 cm in thickness, because of the interaction of turbulence and the boundary layer thickness. The diffusive boundary layer is continually trying to grow, just as the boundary layer of Example 8.5. However, turbulent eddies periodically sweep down and mix a portion of the diffusive boundary layer with the remainder of the fluid. It is this unsteady character of the turbulence near a boundary that is important to the interaction with the diffusive boundary layer.

The eddy diffusion coefficients that we introduced in Chapter 5 were steady quantities, using mean turbulence quantities (e.g., the temporal mean of $u'C'$). This temporal mean character of eddy diffusion coefficients can be misleading in determining the thickness of a diffusive boundary layer because of the importance of unsteady characteristics. We will review some conceptual theories of mass transfer that have been put forward to describe the interaction of the diffusive boundary layer and turbulence.

1. Stagnant Film Theory

The stagnant film theory was developed by Nernst (1904). In this theory, a stagnant film exists on both sides of the interface, as illustrated in Figure 8.8. The thickness of the film is controlled by turbulence and is constant.

With the steady-state situation seen in Figure 8.8, $\partial C/\partial t \Rightarrow 0$, $\partial C/\partial x$ and $\partial C/\partial y \Rightarrow 0$, and $w \Rightarrow 0$. Then, the diffusion equation in each media, as long as turbulence does not penetrate the concentration boundary layer, becomes

$$D_a \frac{\partial^2 C_a}{\partial z^2} = 0 \qquad D_w \frac{\partial^2 C_w}{\partial z^2} = 0 \qquad (8.22)$$

that results in a constant gradient of concentration:

$$\frac{\partial C_a}{\partial z} = \text{const.} \qquad \frac{\partial C_w}{\partial z} = \text{const.} \qquad (8.23)$$

Now, at the interface, the flux in the air phase must equal the flux in the water phase:

$$J_a = -D_a \frac{\partial C_a}{\partial z} \qquad J_w = -D_w \frac{\partial C_w}{\partial z} \qquad (8.24)$$

In addition, we can also apply these equations at the interface:

$$J_a = -k_G(C_a^\infty - C_a^i) \qquad J_w = -k_L(C_w^\infty - C_w^i) \qquad (8.25)$$

where C^∞ is the concentration away from the interface and C^i is the concentration at the interface. Because $\partial C/\partial z$ is constant,

$$\frac{\partial C_a}{\partial z} = \frac{C_a^\infty - C_a^i}{\delta_a} \qquad \frac{\partial C_w}{\partial z} = \frac{C_w^\infty - C_{wa}^i}{\delta_w} \qquad (8.26)$$

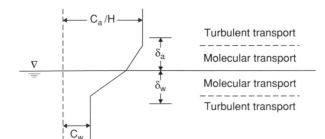

Figure 8.8. Concentration gradient at the interface as assumed by the stagnant film theory.

If we combine equations (8.24) to (8.26), we get

$$K_G = \frac{D_a}{\delta_a} \qquad K_L = \frac{D_w}{\delta_w} \tag{8.27}$$

or

$$Sh_a = \frac{K_G \delta_a}{D_a} = 1 \qquad Sh_w = \frac{K_L \delta_w}{D_w} = 1 \tag{8.28}$$

where Sh is a Sherwood number. Equations (8.27) and (8.28), however, are often valid with one caveat: δ_a and δ_w are not constant, but are a function of time and the diffusion coefficient.

2. Penetration Theory

The penetration theory is attributed to Higbie (1935). In this theory, the fluid in the diffusive boundary layer is periodically removed by eddies. The penetration theory also assumes that the viscous sublayer, for transport of momentum, is thick, relative to the concentration boundary layer, and that each renewal event is complete or extends right down to the interface. The diffusion process is then continually unsteady because of this periodic renewal. This process can be described by a generalization of equation (E8.5.6):

$$\frac{C - C^\infty}{C^i - C^\infty} = \text{erfc}\left(\frac{z}{\sqrt{4Dt}}\right) \tag{8.29}$$

Because the interfacial flux is given as

$$J = -D\frac{\partial C}{\partial z} \tag{8.30}$$

Equations (8.29) and (8.30) result in

$$J = \sqrt{\frac{D}{\pi t}}\, e^{-z^2/4Dt}(C^i - C^\infty) \tag{8.31}$$

or, at $z = 0$,

$$J\,|_{z=0} = \sqrt{\frac{D}{\pi t}}(C^i - C^\infty) \tag{8.32}$$

Combining equation (8.32) with (8.25) provides expressions for both K_L and K_G:

$$K_L = \sqrt{\frac{D_w}{\pi\, t_e}} \qquad K_G = \sqrt{\frac{D_a}{\pi\, t_e}} \tag{8.33}$$

Higbie hypothesized a constant time between events, labeled t_e. After each renewal event, the process of diffusive boundary layer growth would start over again. The diffusive boundary layer growth at one location is illustrated in Figure 8.9. Note that K_L and K_G are now proportional to $(D/t_e)^{1/2}$, instead of D/δ.

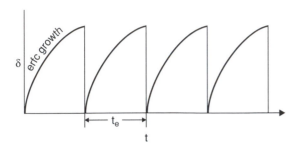

Figure 8.9. Concentration boundary layer growth as visualized by Higbie's penetration theory.

3. Surface Renewal Theory

Developed by Danckwerts (1951), the surface renewal theory states that t_e is not constant, but the renewal of the interface by turbulent eddies will have a period that is represented by a Gaussian distribution with respect to time. Then, K_L or K_G are given by the equation

$$K_L = \sqrt{D_w r} \qquad K_G = \sqrt{D_a r} \tag{8.34}$$

where r is the mean renewal rate (s^{-1}). The assumed thickness of the concentration boundary layer is similar to that given in Figure 8.10.

Instead of determining t_e, in equation (8.33), we must determine r in equation (8.34). Although the difference between Higbie's penetration theory and Danckwerts' surface renewal theory is not great, the fact that a statistical renewal period would have a similar result to a fixed renewal period brought much credibility to Higbie's penetration theory. Equation (8.34) is probably the most used to date, where r is a quantity that must be determined from the analysis of experimental data.

Characterizing Mean Renewal Rate. The mean renewal rate, r, can be characterized using dimensional analysis. Because r has units of inverse time, the characterization

$$r = V^2/\nu \tag{8.35}$$

will result in the appropriate dimensions, where V is some velocity relevant to the problem and ν is kinematic viscosity. If we use the liquid film as an example, then equation (8.34) becomes

$$K_L = \alpha V \sqrt{\frac{D_w}{\nu_w}} \tag{8.36}$$

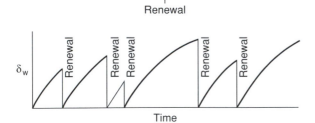

Figure 8.10. Conceptual sketches of surface renewal on the water side of the interface (top) and of the concentration boundary layer thickness at one location over time (bottom), as assumed by the surface renewal theory.

where α is a constant that depends on the application of equation (8.36). This equation may also be written as

$$St\, Sc^{1/2} = \alpha \qquad (8.37)$$

or

$$\frac{Sh}{Re} Sc^{1/2} = \alpha \qquad (8.38)$$

where St is a Stanton number (K_L/V), Sc is a Schmidt number (ν/D_w), Sh is a Sherwood number $(K_L L/D_w)$, and Re is a Reynolds number (VL/ν), and L is some length scale of the flow.

We can also replace V with a shear velocity, $u_* = (\tau/\rho)^{1/2}$, where τ is the interface shear stress and ρ is the density of the fluid. Again, equation (8.34) becomes

$$K_L = \alpha u_* \sqrt{\frac{D_w}{\nu_w}} \qquad (8.39)$$

or

$$St^* = \alpha\, Sc^{1/2} \qquad (8.40)$$

or

$$Sh = \alpha\, Sc^{1/2} Re^* \qquad (8.41)$$

where St^* and Re^* use the shear velocity. Characteristic relations similar to equations (8.39) through (8.41) are commonly used in interfacial mass transfer, because shear velocity is a common representative of the turbulence, and the effect of the turbulence on the concentration boundary layer is what these equations are attempting to represent.

Careful experimentation and analysis by many independent investigators have shown us the following:

a. At fluid–fluid interfaces, $K_L \sim D_w^{1/2}$ and $K_G \sim D_a^{1/2}$ (Daniil and Gulliver, 1991; McCready et al., 1986). The surface renewal theory can be made to fit the transfer data at fluid–fluid interfaces. The exception to this is bubbles with a diameter less than approximately 0.5 mm. Even though there is a fluid on both sides, surface tension causes these small bubbles to behave as though they have a solid–fluid interface. There is also some debate about this 1/2 power relationship at free surfaces exposed to low shear, such as wind–wave flumes at low wind velocity (Jähne et al., 1987) and tanks with surfactants and low turbulence generation (Asher et al., 1996). The difficulty is that these results are influenced by the small facilities used to measure K_L, where surfactants will be more able to restrict free-surface turbulence and the impact on field scale gas transfer has not been demonstrated.

b. At solid–fluid interfaces, $K_L \sim D_w^{2/3}$ and $K_G \sim D_a^{2/3}$ (Campbell and Hanratty, 1982). These data cannot be made to fit the surface renewal theory because the theory gives $K_L \sim D_w^{1/2}$ and $K_G \sim D_a^{1/2}$.

4. Analogy to a Laminar Boundary Layer

Laminar boundary layer theory assumes that a uniform flow ($V = $ constant) approaches a flat plate. A laminar flow region develops near the plate where the thickness of the laminar boundary layer increases with thickness along the plate, as developed in Example 4.2. If we assign δ to be the boundary layer thickness, or the distance from the plate where the velocity is equal to 0.99 times the velocity that approached the plate, and δ_c to be the concentration boundary layer thickness, then we can see that both δ and δ_c are functions of distance, x, from the leading edge, as shown in Figure 8.11.

Example 4.4 gave the following derivation for the boundary layer thickness:

$$\frac{\delta}{x} = 4.64 \ Re_x^{-1/2} \tag{8.42}$$

where $Re_x = x \, V^\infty / \nu$ and V^∞ is the velocity approaching the plate. For a concentration boundary layer on a solid surface (Bird et al., 1960):

$$\frac{\delta_c}{x} = 4.64 \ Re_x^{-1/2} \ Sc^{-1/3} \tag{8.43}$$

If we average the concentration flux over a given length, L, we get

$$Sh = 0.626 \ Re_L^{1/2} \ Sc^{1/3} \tag{8.44}$$

where $Sh = K_B L / D$, $Re_L = L V^\infty / \nu$, and $K_B = $ either K_L or K_G. The analogy is as follows: If equation (8.44) works for laminar flows, why not try it for the mass transfer from a solid surface into a turbulent boundary layer? Only the power on the Reynolds number will be changed, through a fit of the following equation to experimental data:

$$Sh = \alpha_0 \ Re^{\alpha_1} \ Sc^{1/3} \tag{8.45}$$

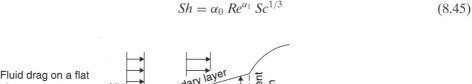

Figure 8.11. Illustration of a boundary layer and a concentration boundary layer on a flat plate. (Adapted from Cussler, 1997.)

where α_0 and α_1 are coefficients to be determined. This is, for the most part, how the characterization equations have been developed in the literature. Table 4.2 provides a number of these equations.

E. Solution of Diffusion Equation Near an Interface

In this section, we will leave the conceptual theories behind and progress to fluid transport with turbulence measurements. This technique was developed by Hanratty and coworkers in the 1970s and 1980s (Campbell and Hanratty, 1982; McCready et al., 1986; Sikar and Hanratty, 1970), with some early free-surface analysis performed by Chen and Scriven (1970). Let us consider a wind blowing over a water surface and the water, in turn, moving above a solid surface. We will be considering the transfer of a volatile compound in which the resistance in the water phase will dominate because of the high resistance to mass transfer in the water phase. In addition, the concentration gradient will be in the water phase, as shown in Figure 8.12. We will assign $z = 0$ at the water surface and at the solid surface.

Now, consider the mass transport equation within roughly $2\delta_c$ of the interface, where there is no turbulent transport because $\delta_c \ll \delta$:

$$\frac{\partial C}{\partial t} + u\frac{\partial C}{\partial x} + v\frac{\partial C}{\partial y} + w\frac{\partial C}{\partial z} = D\left(\frac{\partial^2 C}{\partial x^2} + \frac{\partial^2 C}{\partial y^2} + \frac{\partial^2 C}{\partial z^2}\right) \qquad (8.46)$$

The gradients in the x- and y-directions are probably small when compared with the large gradient in the z-direction. This will be sufficient to overcome the fact that u and v are greater than w. Thus,

$$w\frac{\partial C}{\partial z} \gg u\frac{\partial C}{\partial x}, \quad v\frac{\partial C}{\partial y} \qquad (8.47)$$

and

$$\frac{\partial^2 C}{\partial z^2} \gg \frac{\partial^2 C}{\partial x^2}, \quad \frac{\partial^2 C}{\partial y^2} \qquad (8.48)$$

Then, equation (8.46) becomes

$$\frac{\partial C}{\partial t} + w\frac{\partial C}{\partial z} = D\frac{\partial^2 C}{\partial z^2} \qquad (8.49)$$

What we have done with equation (8.49) is convert C and w from functions of x, y, z, and t to functions of z and t. In equation (8.49), we then have the parameters $w(z,t)$, $C(z,t)$, and D. The only unknowns are the dependent variable, C, and w, as D can be

Figure 8.12. Illustration of volatile compound transport from a solid surface and from the water into the air.

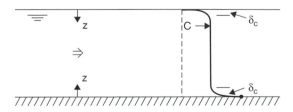

determined either from measurements or from the predictive techniques described in Chapter 3. If we can somehow determine $w(z, t)$, we can numerically solve for $C(z, t)$.

In addition, we know these relationships for mass flux:

$$J = -D\frac{\partial C}{\partial z}\Big|_{z=0} = K_L(C^\infty - C^i) \tag{8.50}$$

Thus, if we can determine $C(z, t)$, we can also apply equation (8.50) to find $K_L(t)$, and

$$\overline{K_L} = \frac{1}{\Delta t} \int_{\Delta t} K_L(t)dt \tag{8.51}$$

We now have the following recipe to determine $K_L(t)$ and $\overline{K_L}$:

1. Given D, measure $w(z,t)$ within $2\delta_c$ of the interface.
2. For given boundary conditions, such as $C^I = C^*$ and $C^\infty = 0$, solve equation (8.49) numerically for $C(z,t)$ and $\partial C/\partial z(t)$ at $z = 0$.
3. Solve equation (8.50) for $K_L(t)$.
4. Solve equation (8.51) for $\overline{K}_L$.

Note that our recipe does not require measurement of any concentration – that is because $K_L = K_L(D, \text{turbulence})$. K_L is *not* a function of concentration. It is thus logical that we do not have to measure concentration to measure K_L, as long as we know the fundamental relation between turbulence and mass transfer. Of course, there are difficulties with this four-step recipe. Primarily, item 1: "Measure $w(z,t)$ within $2\delta_c$ of the interface." In water, a fairly large concentration boundary layer would be $100\,\mu m$. How are we going to get a profile of velocity measurements to determine $w(z,t)$ within $200\,\mu m$ of the interface, when the measuring volumes of most instruments are greater than $200\,\mu m$? This is the problem that Thomas Hanratty and coworkers recognized and solved.

Hanratty et al.'s Solution to Recipe Item 1. Recipe item 1 will have a set of substeps that must be undertaken to find a solution to this measurement problem. They are:

1a. Assume that $w(z,t)$ may be separated into two individual functions of time and distance, respectively, or

$$w(z, t) = \beta(t)g(z) \tag{8.52}$$

where $\beta(t)$ is Hanratty's beta, representing the driving turbulence outside of the concentration boundary layer, δ_c, and $g(z)$ represents the stretching, shrinking, and overall dissipation of the turbulent eddies as they approach the interface. These processes are illustrated in Figure 8.13.

1b. Expand $g(z)$ from $z = 0$ to $z = \delta_c$, as in equation (8.53):

$$g(z) = g_0 + g_1 z + g_2 z^2 + \cdots \tag{8.53}$$

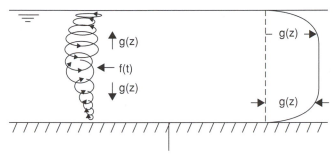

Figure 8.13. Illustration of the two processes represented by $f(t)$ and $g(z)$ in the interface turbulence decomposition of Hanratty and others.

1c. Apply continuity at the interface, shown in equation (8.54):

$$\frac{\partial w}{\partial z} = \frac{\partial g}{\partial z}\beta(t) \tag{8.54}$$

We now need to determine $\partial u/\partial x$ and $\partial v/\partial y$ in equation (8.54). This will be accomplished in steps 1d, 1e, and 1f.

1d. Expand u and v about the interface ($z = 0$). Then,

$$u(x, y, z, t) = \sum_{k=0}^{\infty} \sum_{j=0}^{\infty} \sum_{i=0}^{\infty} \phi_{i,j,k} x^i y^j z^k \tag{8.55}$$

and

$$v(x, y, z, t) = \sum_{k=0}^{\infty} \sum_{j=0}^{\infty} \sum_{i=0}^{\infty} \eta_{i,j,k} x^i y^j z^k \tag{8.56}$$

where φ and η are both some function of time. Equations (8.55) and (8.56) do not truly need to be evaluated to $(i, j, k) = \infty$, because we are actually very close to $z = 0$, and the higher order terms will not be significant. Applying the summations for 0 and 1, for example, is sufficient, and results in four terms in equations (8.55) and (8.56).

1e. Apply the boundary conditions at $z = 0$. For a free interface, these are $w = 0$, $\partial u/\partial z = 0$, and $\partial v/\partial z = 0$. At a fixed interface, the boundary conditions are $w = 0$, $u = 0$, and $v = 0$.

1f. Solve equations (8.53) through (8.56) for $g(z)$. The result of this operation is different for a *free interface*, given in equation (8.57),

$$w = \beta(t)z \tag{8.57}$$

than the result for a *fixed interface*, given in equation (8.58):

$$w = \beta(t)z^2 \tag{8.58}$$

The difference between a free interface (equation (8.57)) and a fixed interface (equation (8.58)) will eventually result in a different dependence of K_L on diffusion coefficient.

Campbell and Hanratty (1982) used Lau's (1980) measurements with some special optics on a laser Doppler velocimetry system to calculate $\beta(t)$ near a fixed interface, in this case, the inside of a clear pipe. They determined $w(z,t)$ from equation (8.52), and solved equations (8.49) and (8.50) numerically for $K_L(t)$. Finally, they applied equation (8.51) to determine K_L, which has been the goal all along. The end results (K_L) may then be related to the other, independent parameters that are important to the transfer process, such as diffusivity, viscosity, and turbulence parameters. Campbell and Hanratty performed this operation and found the following correlation:

$$K_L = u_* F[\beta(t)] Sc^{-7/10} \tag{8.59}$$

or

$$Sh = F[\beta(t)] Re_* Sc^{0.3} \tag{8.60}$$

where F is a function that uses the frequency spectra of $\beta(t)$.

$$F[\beta(t)] = 0.237 S_{\beta m}^{+0.21} \tag{8.61}$$

where $S_{\beta m}^{+} = S_{\beta m} v / u_*^2$ and $S_{\beta m}$ is the maximum value of the frequency spectrum of β with units of time^{-1}. Note that equation (8.60) is proportional to $Sc^{0.3}$, rather that $Sc^{1/3}$ for the fixed interface. $Sc^{1/3}$ was based on an *analogy* to a laminar boundary layer, which worked fairly well for fixed interfaces but was not necessarily correct for a turbulent boundary layer. The correlation of Campbell and Hanratty (1982) was based on actual data and on numerical simulations over the range of relevant conditions. It is likely that the 0.3 power relationship is the correct one.

So, now that we have the fixed interface fairly well described, what about the free interface? This actually presents another problem for the techniques of Campbell and Hanratty. The $z = 0$ axis is *on* the water surface. If the water surface moves up and down, the axis moves with it, and is actually a moving coordinate system. It is difficult to get a water surface, exposed to turbulence, that does not move more than 100 μm.

McCready et al. (1986) assumed that the turbulence that was measured by Campbell and Hanratty near a fixed interface could also be applied to a free interface (i.e., $f(t)$ could be the same); it was only $g(z)$ that would be different. Thus, the same velocity measurements were applied to a free surface. The results of the numerical analysis were quite different and are given in equations (8.62) and (8.63).

$$K_L = (0.5 \beta_{\mathrm{rms}} D)^{1/2} \tag{8.62}$$

or

$$Sh_* = 0.71 \ Sc^{1/2} \beta^{+1/2} Re_* \tag{8.63}$$

where $\beta_{\mathrm{rms}} = [\overline{\beta(t)^2}]^{1/2}$ is the root mean square value of β, and $\beta^+ = \beta_{\mathrm{rms}} v / u_*^2$. McCready et al. used a linear approach applied to equation (8.49) to explain the parameterization differences between equations (8.59) through (8.61) and equation (8.63): if all of the scales of β are important to gas transfer, then β_{rms} is the independent parameter to characterize K_L. If only the larger scales of β are important, then $S_{\beta m}$ is the parameter that should be in the characterization.

From equation (8.62), we can see that we also have a "measurement" of the surface renewal rate, r:

$$r = 0.5\beta_{rms} \tag{8.64}$$

The penetration and surface renewal theories started out as conceptual, in that they were visualized to occur as such by individual theorists. These theories appeared to work successfully for a free interface, such as the air–water interface, but not for a fixed interface, such as solid–water. Now, the explanation is before us in equation (8.64). Surface renewal is a fairly accurate representation of Hanratty's β at a free surface, and therefore can be seen to give representative results. It is Hanratty's β that we really should be measuring, and it happens that the mean surface renewal rate is a good representation of Hanratty's β at a free surface.

Attempts were made to visualize and measure these surface renewal events on a free surface (Davies and Khan, 1965; Gulliver and Halverson, 1989; Komori et al., 1982; Rashidi and Banerjee, 1988), but measurements were still concentrating on a parameter from a conceptual theory, with no ability to distinguish if it was truly related to K_L. Tamburrino and Gulliver (1994, 2002) developed a means of visualizing and measuring Hanratty's β through streakline imaging or particle tracking velocimetry on a water surface that is close to horizontal to the camera. Applying continuity to equation (8.57) at the free surface:

$$\beta = -\left(\frac{\partial u}{\partial x} + \frac{\partial v}{\partial y}\right) \tag{8.65}$$

Equation (8.65) tells us that if we can measure the two-dimensional divergence on a free surface, then we are also measuring Hanratty's β. A sample of these types of results, given in Figure 8.14, indicates that the scale of Hanratty's β is a fraction of that associated with lower frequency events. This means that virtually all of the measurements of surface renewal have proven to be measuring something else, because they concentrated on the low-frequency events. On a laboratory scale, we are currently able to take measurements of two-dimensional divergence that can be used to describe the liquid film coefficient directly through equations like equation (8.62).

Other measurements of Hanratty's β have been made or inferred from various techniques, including a hot film probe just under the water surface (Brumley and Jirka, 1987), particle image velocimetry in a vertical laser sheet leading up to the water surface with a florescent dye to indicate water surface location accurately (Law and Khoo, 2002) and PIV on the water surface (McKenna and McGillis, 2004; Orlins and Gulliver, 2002). The measurements of Law and Khoo (2002) are especially interesting because the following relationship was developed from experiments on both a jet-stirred tank and a wind–wave channel:

$$K_L = (0.05 D\beta_{rms})^{1/2} \tag{8.66}$$

or

$$Sh = 0.22 \, Sc^{1/2}\beta^{+1/2} \, Re \tag{8.67}$$

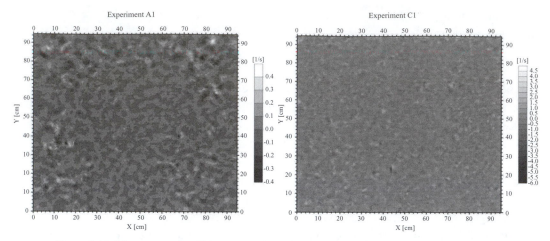

Figure 8.14. Measurements of Hanratty's β in a 95-cm tank stirred by jets emanating from the bottom. *Left* and *right* images are at depths of 44 cm and 22 cm, respectively (Adapted from Tamburrino and Aravena, 2000).

where any convenient turbulence length and velocity scale may be used to make the parameters nondimensional, as long as it is similar for each dimensionless number. The similarities between equations (8.66) and (8.62) are striking, as are the differences. The two equations result in a similar form for the K_L relationship. However, McCready et al. ended up with a coefficient for the equivalent renewal rate of 0.5, while Law and Khoo ended up with 0.05.

In contrast, Tamburrino and Gulliver (2002) related liquid film coefficient from Gulliver and Halverson (1989) and Lau (1975) to their measurements of Hanratty's β for open-channel flows in a flume. They could not get a $K_L \sim \beta_{\text{rms}}^{1/2}$ relationship. Instead, their result was the following:

$$Sh = Sc^{-1/2} \left(\beta^+ Pe_*^2 \right)^{0.645} \tag{8.68}$$

where $Pe_* = u_* H / v$. The exponent of 0.645 in equation (8.68) is a problem because the relation for K_L now depends on the turbulence velocity and length scales chosen, which should be avoided with this fundamental approach. Law and Khoo's experiment had higher values of β, which may explain the difference. Tamburrino and Gulliver's hypothesis was that, because the turbulence originated from the channel bottom and went through transitions as it moved to the water surface, the larger β scales had the primary influence on gas transfer coefficient. They computed the frequency spectra of β, verified that their supposition was true, and found the following relationship that gave a dimensionally correct relationship:

$$K_L = (0.058 \, S_{\beta m} D)^{1/2} \tag{8.69}$$

or

$$Sh = 0.24 \, Sc^{1/2} S_{\beta m}^+ Re \tag{8.70}$$

Again, any turbulence velocity and length scale are sufficient to use in these dimensionless parameters, as long as they are used in all parameters. Tamburrino and Gulliver used the bottom shear velocity and channel depth. Equation (8.69) provides a measure of surface renewal rate:

$$r = 0.058 \, S_{\beta m} \qquad (8.71)$$

So, this fundamental approach to determining gas transfer coefficient is still sorting out which flows have all scales contribute to gas transfer coefficient, such that $K_L \sim \beta_{rms}^{1/2}$, and which flows have only the larger scales of β contribution, such that $K_l \sim S_{\beta m}^{1/2}$. It will be some time before the approach is taken to the field.

F. Gas Film Coefficient

The gas film coefficient is dependent on turbulence in the boundary layer over the water body. Table 4.1 provides Schmidt and Prandtl numbers for air and water. In water, Schmidt and Prandtl numbers on the order of 1,000 and 10, respectively, results in the entire concentration boundary layer being inside of the laminar sublayer of the momentum boundary layer. In air, both the Schmidt and Prandtl numbers are on the order of 1. This means that the analogy between momentum, heat, and mass transport is more precise for air than for water, and the techniques applied to determine momentum transport away from an interface may be more applicable to heat and mass transport in air than they are to the liquid side of the interface.

There are two types of convection in an air boundary layer: (1) forced convection created by air movement across the water body and (2) free convection created by a difference in density between the air in contact with the water surface and the ambient air. If the water body is warmer than the surrounding air, free convection will occur. The combination of these two processes is illustrated in Figure 8.15.

The gas film coefficient due to *free convection* (Figure 8.15a) is described by the relation of Shulyakovskyi (1969):

$$K_G = 0.14 \left(\frac{g \alpha_a^2 \beta_T \Delta \theta_v}{v_a} \right)^{1/3} \qquad \text{Positive } \Delta \theta_v \qquad (8.72)$$

where g is the acceleration of gravity, α_a is the thermal diffusion coefficient in air, $\beta_T(1/°)$ is the coefficient of thermal expansion for moist air, v_a is the kinematic viscosity of air, and $\Delta \theta_v$ is the virtual temperature difference between air at the water surface and the ambient air, representing a density difference. Virtual temperature is the density of dry air at otherwise similar conditions. The relation for $\Delta \theta_v$ is:

$$\Delta \theta_v = \| T_s \left(1 + 0.378 p_{vs} / P_a \right) - T_a \left(1 + 0.378 p_{va} / P_a \right), 0 \| \qquad (8.73)$$

where $\|\ \|$ stands for the greater of the two terms in brackets; T_s and T_a are the temperatures of the water surface and ambient air, respectively; p_{vs} and p_{va} are the vapor pressures of saturated air on the water surface and ambient air, respectively; and P_a is the air pressure.

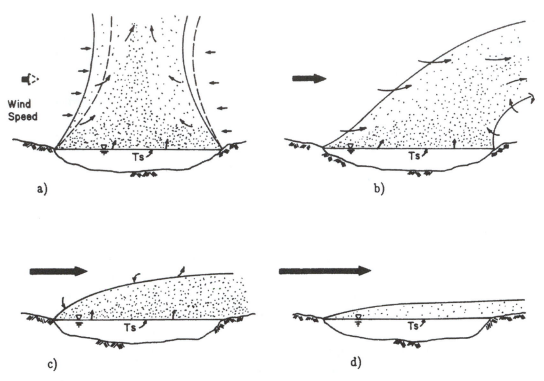

Figure 8.15. Free and forced convection regimes: (a) free convective plume, where $T_s > T_{air}$; (b) convective plume deflected by wind; (c) unstable boundary layer; and (d) stable or neutral boundary layer, where $T_s < T_{air}$ (Adams et al., 1990).

Saturation vapor pressure at any air temperature may be computed by the Magnus-Tetons formula:

$$p_{vs}(mb) = 6.11 \exp\left\{\frac{17.3[T(^{\circ}K) - 273]}{T(^{\circ}K) - 35.9}\right\} \tag{8.74}$$

and atmospheric vapor pressure can be computed from the saturation vapor pressure and relative humidity, RH:

$$p_{va} = \frac{RH}{100}p_{vs} \tag{8.75}$$

Because values of α_a and v_a can be found from tables such as those in the *Handbook of Chemistry and Physics* (CRC, 2005), all that remains is to determine β_T before we can estimate K_G from free convection. The equation

$$\beta_T = \frac{1}{T_{av}(^{\circ}C) + 273} \tag{8.76}$$

will provide us with this parameter, where $T_{av} = T_a(1 + 0.378p_{va}/P_a)$ is the virtual temperature of the air.

Adams et al. (1990) have investigated the application of the free convection relationships for the cooling water pond at the Savannah River thermal power plant and found that free convection should dominate at wind speeds (at 2 m height) of less than $0.5 - 1$ m/s. At wind speeds greater than 1 to 1.5 m/s, the evaporation regime is that of an unstable forced convective layer, illustrated in Figure 8.15c. Thus, the deflected plume, Figure 8.15b, is a transition regime that will be "smoothed over" for evaporation from water bodies. A stable boundary layer occurs when equation (8.73) is equal to zero.

The *unstable boundary layer* is a flow regime that exists at wind speeds greater than approximately 1.5 m/s when $\Delta\theta_v$ is positive. This is a commonly occurring regime in environmental flows. We will start by assuming the common logarithmic velocity profile in the air boundary layer, altered by a stability function (Brutsaert, 1982) and assign $J_m = \rho u_*^2$ to be the flux of momentum at height z. Then, the momentum/unit volume difference between heights z and $z = 0$ is

$$\rho\left[u(z) - u(z = 0)\right] = \frac{J_m}{\kappa u_*}\left[\ln\left(\frac{z}{z_{0m}}\right) - \psi_m\right] = \frac{u_*}{\kappa}\left[\ln\left(\frac{z}{z_{0m}}\right) - \psi_m\right] \qquad (8.77)$$

where $\kappa \approx 0.4$ is von Karmon's constant; z_0 is dynamic roughness, which is a fraction of the true surface roughness (Turner, 1994); and ψ_m is a momentum-flux stability function that is nonzero with stable and unstable boundary layers. For unstable boundary layers, ψ_m is given by (Brutsaert, 1982):

$$\psi_m = \ln\left[\frac{1 + (1 - 16z/L)^{1/2}}{1 + (1 - 16z_{0m}/L)^{1/2}}\right] + 2\ln\left[\frac{1 + (1 - 16z/L)^{1/4}}{1 + (1 - 16z_{0m}/L)^{1/4}}\right]$$
$$- 2\arctan\left[(1 - 16z/L)^{1/4}\right] + 2\arctan\left[(1 - 16z_{0m}/L)^{1/4}\right] \qquad (8.78)$$

and L is Obukhov's stability length,

$$L = -\frac{u_*^3}{\kappa B} \qquad (8.79)$$

where B is the buoyancy flux per unit area, expressed as

$$B = \frac{g\beta_T(\varphi_c + 0.07\varphi_e)}{\rho C_p} \qquad (8.80)$$

where ρ is the density of air, C_p is the specific heat capacity of dry air, φ_c is the heat flux across the water surface by conduction, and φ_e is the evaporative heat flux from the water surface. Equation (8.78) results from the integration of a relatively simple equation curve fit to field data. The integration results in this complicated equation. For a smooth surface, $u_* z_{0m}/v$ and $u_* z_{0A}/D_A$ are both equal to 9.6. At most wind velocities and wind fetches, however, the water surface is not hydrodynamically smooth, and another relation is needed.

The equation for mass flux of compound A, J_A, and mass of A per unit volume will be written by analogy to equation (8.77):

$$C_A(z) - C_A(z = 0) = \frac{J_A}{\kappa u_*}\left[\ln\left(\frac{z}{z_{0A}}\right) - \psi_A\right] \qquad (8.81)$$

where ψ_A is given by (Brutsaert, 1982):

$$\psi_A = 2\ln\left[\frac{1 + (1 - 16z/L)^{1/2}}{1 + (1 - 16z_{0A}/L)^{1/2}}\right] \qquad (8.82)$$

If equations (8.77) and (8.81) are combined to eliminate shear velocity, an equation can be developed for the flux of compound A:

$$J_A = \frac{\kappa^2 \left[C_A(z) - C_A(z=0)\right] \left[u(z) - u(z=0)\right]}{\left[\ln\left(\dfrac{z}{z_{0A}}\right) - \psi_A\right]\left[\ln\left(\dfrac{z}{z_{0m}}\right) - \psi_m\right]} \tag{8.83}$$

Equation (8.83) provides us with an expression for the gas film coefficient for an atmospheric boundary layer:

$$K_G = \frac{\kappa^2 \left[u(z) - u(z=0)\right]}{\left[\ln\left(\dfrac{z}{z_{0A}}\right) - \psi_A\right]\left[\ln\left(\dfrac{z}{z_{0m}}\right) - \psi_m\right]} \tag{8.84}$$

The gas film coefficient and gas flux rate can be estimated through iteration on equations (8.77) through (8.84), if φ_c and φ_e are known, and a good relationship for z_{0m} and z_{0A} is known. Evaporative heat flux can be determined by applying equation (8.85) to water vapor and using the relationship

$$\varphi_e = L_v J_{H_2O} \tag{8.85}$$

where L_v is the latent heat of vaporization, given as

$$L_v(\text{cal/g}) = 597 - 0.571 T_s(^\circ\text{C}) \tag{8.86}$$

where T_s is water surface temperature. Then, Bowen's ratio (Bowen, 1926) may be used to compute φ_c:

$$\varphi_c = 0.61 P_a(\text{bars}) L_e K_G \left[T_s - T(z)\right] \tag{8.87}$$

where P_a is air pressure. The application of these equations to determine gas film coefficient will be demonstrated in Example 8.6.

EXAMPLE 8.6: *Computation of gas film coefficient over a water body*

Long lake, with a fetch of 10 km, is exposed to various wind speeds. On a cold day in fall, the water temperature at 10°C has not cooled yet, but the air temperature is at −5°C. The relative humidity is 100%. Compute the gas film coefficient for water vapor at various wind fetch lengths for wind speeds at 2 m height, between 2 and 20 m/s.

For these conditions, we have the following water vapor concentrations:

$C_{H_2O}(z=0) = p_{vs}\rho/P_a = 13\,\text{g/m}^3$, assuming that $P_a = 1.013$ bars, and,
$C_{H_2O}(z) = 4\,\text{g/m}^3$

In addition, we need to find a parametric relationship for z_0, which will be equal to z_{0H_2O} at velocities above 2 m/s, because the boundary of the water surface will

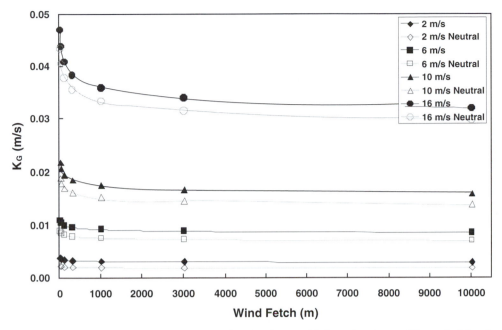

Figure E8.6.1. Computations of gas film coefficient for the unstable boundary layer of Example 8.6 and for a neutral boundary layer that is otherwise similar.

be rough. Gulliver and Song (1986) combined the relationships of Hsu (1974) and Mitsuyatsu (1968) for gravity waves to get

$$\frac{z_0 g}{u_*^2} = 0.00942\pi \left(\frac{gF}{u_*^2}\right) \tag{E8.6.1}$$

and the relations of Hsu and Long and Hwang (1976) for capillary waves to get

$$\frac{z_0 g}{u_*^2} = 0.0219\pi \left(\frac{gF}{u_*^2}\right)^{1/2} \left(\frac{\sigma}{\rho g F}\right)^{1/3} \tag{E8.6.2}$$

where σ is the surface tension of clean water exposed to air. The transition is presumed to occur when z_0 from equation E8.6.2 becomes greater than that of equation (E8.6.1). Finally, Wu (1975) gives

$$u(z = 0) = 0.55u_* \sqrt{\frac{\rho}{\rho_w}} \tag{E8.6.3}$$

where ρ_w is the water density.

We will assume that $z_{0m} = z_{0A} = z_0$, which is generally true except for a smooth water surface. When we use equations (E8.6.1) through (E8.6.3), along with the equations provided above (8.77) through (8.87) to compute K_G for water vapor versus fetch for different wind velocities, the result is that provided in Figure E8.6.1. Iteration was required on the entire set of equations. Also provided in Figure E8.6.1 are similar results for neutral conditions, with $\psi_m = \psi_{H_2O} = 0$. The gas film coefficient decreases with fetch length as z_0 decreases due to an increase in wave velocity. Even though the water surface is not moving quickly, as equation (E8.6.3) can be used to

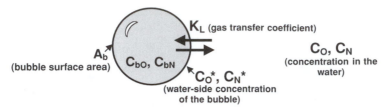

Figure 8.16. Parameters important to the bubble–water gas transfer.

demonstrate, the waves that provide the roughness are moving faster as the wind fetch increases.

G. Bubble–Water Gas Transfer

Bubbles complicate the mass transfer process because the concentration of gases in the bubble is not constant with time. Instead of being exposed to the atmosphere – which is assumed to be a large container such that ambient concentrations do not change in the time of interest – a bubble volume is more limited, and the concentration of the various compounds can change due to mass flux or due to a change in pressure. This means that, for a volatile compound (for a gas), the equilibrium concentration at the water surface is not constant.

We will perform a mass balance on oxygen, which is commonly of interest in bubble–water gas transfer and is illustrated in Figure 8.16. Bubbles are often placed into the water because of their high surface area and tendency to generate their own turbulence. A similar mass balance may be performed on any compound in the water or the gas in the bubbles. Air is mostly oxygen and nitrogen, as given in Table 8.3, with a smaller concentration of argon and traces of other compounds. Humid air, such as air bubbles, also needs to incorporate water vapor using equation (8.74) and the relation

$$C_{H_2O}(\%) = \frac{p_{vs}}{P_a} \qquad (8.88)$$

Table 8.3: *Composition of dry air (CRC, 2005)*

Gas	Symbol	% by volume	% by mass	Molecular weight
Nitrogen	N_2	78.08	75.47	28.01
Oxygen	O_2	20.95	23.20	32.00
Argon	Ar	0.93	1.28	39.95
Carbon dioxide	CO_2	0.038	0.0590	44.01
Neon	Ne	0.0018	0.0012	20.18
Helium	He	0.0005	0.00007	4.00
Krypton	Kr	0.0001	0.0003	83.80
Hydrogen	H_2	0.00005	Negligible	2.02
Xenon	Xe	8.7×10^{-6}	0.00004	131.30

Then, to keep the total at 100%, other gas concentrations need to be adjusted accordingly.

We will conserve both nitrogen and oxygen gas. Often, argon is considered to be similar to nitrogen because both are nonreactive and have close to the same diffusion coefficient. Because both gases are volatile, there is no significant resistance on the gas side of the interface, and

$$C_{bO} = H_O C_O^* \tag{8.89a}$$

$$C_{bN} = H_N C_N^* \tag{8.89b}$$

Then, the change of concentration in the water phase is given by

$$V \frac{dC_O}{dt} = K_{LO} A_b (C_O^* - C_O) + K_{LSO} A_s (C_{sO} - C_O) \tag{8.90a}$$

and

$$V \frac{dC_N}{dt} = K_{LN} A_b (C_N^* - C_N) + K_{LSN} A_s (C_{SN} - C_N) \tag{8.90b}$$

where V is the volume of water; subscripts N and O indicate nitrogen and oxygen compounds, respectively; K_L is the liquid film coefficient; K_{LS} is the liquid film coefficient of the water exposed to the atmosphere; A_b is the bubble surface area; A_s is the surface area exposed to the atmosphere; C is the water-side concentration; C^* is the concentration of the water in equilibrium with the bubble; and C_S is the saturation concentration of water exposed to the atmosphere. For bubbles greater than approximately 0.5 mm in diameter, the bubble surface acts similar to a free surface because of the toroidal circulation of gas inside of the bubble, as illustrated in Figure 8.17. Then, equation (8.62) applies:

$$K_{LSN} = \left(\frac{D_N}{D_O} \right)^{1/2} K_{LSO} \tag{8.91a}$$

$$K_{LN} = \left(\frac{D_N}{D_O} \right)^{1/2} K_{LO} \tag{8.91b}$$

where D is the diffusion coefficient of the compound in water. For bubbles less than approximately 0.5 mm in diameter, the bubble acts similar to a solid and equation (8.59) applies. This gives

$$K_{LN} = \left(\frac{D_N}{D_O} \right)^{0.3} K_{LO} \tag{8.92}$$

The concentration inside of the bubble must also be considered:

$$V_b \frac{dC_{bO}}{dt} = K_{LO} (C_O^* - C_O) \tag{8.93a}$$

$$V_b \frac{dC_{bN}}{dt} = K_{LN} (C_N^* - C_N) \tag{8.93b}$$

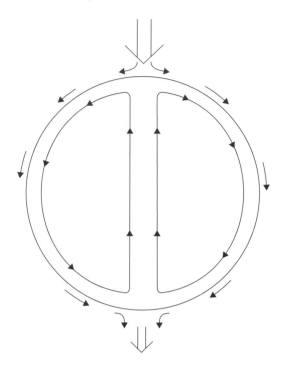

Figure 8.17. Circulation of gas inside of a bubble. The coordinate system is moving with the bubble.

where V_b is the volume of the bubble and C_b is the concentration of the respective compound in the bubble. Because of the difference in pressure that a bubble experiences, it is easier to keep track of the mole ratio, y_m:

$$y_m = \frac{C_{bO}}{C_{bN}} \tag{8.94}$$

where concentration is given in moles/volume, and

$$\frac{dy_m}{dt} = \frac{\partial y_m}{\partial t} + v_r \frac{\partial y_m}{\partial z} \tag{8.95}$$

where v_r is the rise velocity of the bubble. The $\partial y_m / \partial t$ term can be visualized as the change of y_m over time at one location, with many bubbles streaming by. This term is generally small compared with the last term in equation (8.95) and will be assumed negligible for the computation of y_m, as long as accurate values of C_O, C_N, etc., are known. We will therefore concentrate on the $\partial y_m / \partial z$ term. Thus,

$$\frac{\partial y_m}{\partial z} = \frac{\partial (C_{bO}/C_{bN})}{\partial z} = \frac{1}{C_{bN}} \frac{\partial C_{bO}}{\partial z} - \frac{C_{bO}}{C_{bN}^2} \frac{\partial C_{bN}}{\partial z} \tag{8.96}$$

To determine C_{bO} and C_{bN}, we will return to the mass flux:

$$v_r V_b \frac{\partial C_{bO}}{\partial z} = K_{LO} (C_O^* - C_O) \tag{8.97a}$$

$$v_r V_b \frac{\partial C_{bN}}{\partial z} = K_{LN} (C_N^* - C_N) \tag{8.97b}$$

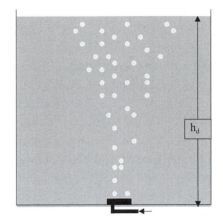

Figure 8.18. Illustration of a bubble plume in a tank.

The terms v_r and V_b are terms that must be estimated. However, the multiplicative pair, $v_r V_b$, can be converted through analysis of a rising bubble plume (McWhirter and Hutter, 1989), as shown in Figure 8.18. The residence time of bubbles in the tank, t_r, is given by $t_r = V_b/Q_a$, where Q_a is the discharge of air into the tank. In addition, $t_r = h_d/v_r$, where h_d is the depth of water in the tank. Then,

$$t_r = \frac{V_b}{Q_a} = \frac{h_d}{v_r} \qquad (8.98)$$

which means that

$$V_b v_r = Q_a h_d \qquad (8.99)$$

Thus, the unknown term in equation (8.97), $V_b v_r$, has been replaced with the known terms, $Q_a h_d$.

Substituting equation (8.99) into (8.97) gives

$$\frac{\partial C_{bO}}{\partial z} = \frac{K_{LO} A_b}{Q_a h_d} (C_O^* - C_O) \qquad (8.100a)$$

$$\frac{\partial C_{bN}}{\partial z} = \frac{K_{LN} A_b}{Q_a h_d} (C_N^* - C_N) \qquad (8.100b)$$

Now, substituting equations (8.89), (8.91), and (8.100) into (8.96) gives

$$\frac{\partial y_m}{\partial z} = \frac{-K_{LO} A_b}{h_d Q_a} \left[\frac{1}{H_N C_N^*} (C_O^* - C_O) - \left(\frac{D_N}{D_O} \right)^{1/2} \frac{H_O C_O^*}{(H_N C_N^*)^2} (C_N^* - C_N) \right] \qquad (8.101)$$

These equations are finalized by determining the equilibrium concentration between the bubble and the water on the edge of the bubble:

$$C_O^* = C_{SO} \frac{P_a - p_v + [\rho_L g(h_d - z)]}{P_{atm} - p_v} \frac{y_m}{0.208(1 + y_m)} \qquad (8.102a)$$

$$C_N^* = C_{SN} \frac{P_a - p_v + [\rho_L g(h_d - z)]}{P_{atm} - p_v} \frac{1}{0.208(1 + y_m)} \qquad (8.102b)$$

where ρ_L is the density of the liquid, P_a is the pressure of the local atmosphere, p_v is the vapor pressure, and P_{atm} is the pressure of the standard atmosphere. Equation (8.102) changes with depth as z and y_m change with depth. The boundary condition of equations (8.101) and (8.102) is the gas molar ratio at the sparger, which is 0.266 for air and the initial concentrations C_O and C_N in the liquid. The equations are then solved from $z = 0$ to $z = h_d$ at one moment in time and used to compute C_O and C_N in equation (8.90) for the next time step.

We can see that the equations and solution technique have an added degree of complexity for bubble–water gas transfer, primarily because of the variation of pressure and because the gas control volume cannot be considered large. These are, however, simply the mass conservation equations, which should not be considered difficult, only cumbersome. Any reduction of these equations is an assumption, which would need to be justified for the particular application. If transfer of a trace gas is of interest, then similar equations for the trace gas would need to be added to those provided above.

We also need to develop the theories for liquid film coefficient to use in the aforementioned equations. For drops that are close to spherical, without separation, Levich (1962) assumed that the concentration boundary layer developed as the bubble interface moved from the top to the bottom of a spherical bubble. Then, it is possible to use the concepts applied in Section 8.C and some relations for the streamlines around a bubble to determine K_L:

$$Sh = \left(\frac{4}{\pi}Pe\right)^{1/2} \tag{8.103}$$

where $Sh = K_L d_b / D$, $Pe = v_r d_b / D$, and d_b is the bubble diameter. Equation (8.103) should apply as long as the bubble is spherical and the interface moves with the rise of the bubble. For bubbles where the surface acts as a solid surface (does not move), and is spherical, Friedlander (1961) developed the equation

$$Sh = 0.99 Pe^{1/3} \tag{8.104}$$

that applies to very small bubbles. In general, most of the gas transfer is across bubbles that are not spherical, because of turbulence generated in the bubble rise field, so variations on equations (8.103) and (8.104) have been proposed, as given in Table 4.2. These equations, however, do provide a good basis from which to develop characterization equations.

H. Interfacial Transfer with Reaction

There are a number of environmentally significant compounds that undergo a reaction while moving through the water-side, concentration boundary layer, such that the flux rate is altered. If the flux rate is altered, then the apparent rate coefficient is also affected. Typical examples would be the compounds that react with

Table 8.4: *Environmentally significant compounds that react with water*

								$k\,(s^{-1})$
CO_2	+	H_2O	$\Rightarrow$	H_2CO_3				3.7×10^{-2}
Cl_2	+	H_2O	$\Rightarrow$	$HOCl$	+	H^+	$+ Cl^-$	28
H_2S	+	H_2O	$\Rightarrow$	HS^-	+	H_3O^+		$4. \times 10^2$
NH_3	+	H_2O	$\Rightarrow$	$NH_4{}^+$	+	OH^-		
CH_3COOH	+	H_2O	$\Rightarrow$	CH_3COO^-	+	H_3O^+		$1. \times 10^5$
(Methyl formate)								

water, such as the reactions given in Table 8.4. These reactions are all essentially *irreversible*.

EXAMPLE 8.7: *Absorption rate of carbon dioxide by the oceans*

Carbon dioxide is one of the major global warming gases. The ocean acts as a reservoir for carbon dioxide and therefore will slow the effects of this gas on global warming. How is this air–water transfer rate dependent on the rate of reaction of carbon dioxide with liquid water to form H_2CO_3?

Carbon dioxide has a source/sink rate of $-k\,C$, where C is the water concentration of carbon dioxide. The concentration profile in the concentration boundary layer would be steeper near the interface, as illustrated in Figure E8.7.1.

Thus, the diffusion equation within the water concentration boundary layer becomes

$$\frac{\partial C}{\partial t} = D\frac{\partial^2 C}{\partial z^2} - kC \tag{E8.7.1}$$

with boundary conditions:

1. At $t = 0^+$, $z = 0$, $C = C_0$
2. At $t = 0$, $z > 0$, $C = 0$

The solution to equation (E8.7.1) has been developed by Danckwerts (1970):

$$\frac{C}{C_0} = \frac{1}{2}\exp\left(-z\sqrt{\frac{k}{D}}\right)\operatorname{erfc}\left(\frac{z}{2\sqrt{tD}} - \sqrt{kt}\right) + \frac{1}{2}\exp\left(z\sqrt{\frac{k}{D}}\right)\operatorname{erfc}\left(\frac{z}{2\sqrt{tD}} + \sqrt{kt}\right) \tag{E8.7.2}$$

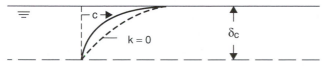

Figure E8.7.1. Concentration profile with a first-order, irreversible degradation reaction (*solid line*) and an equivalent concentration profile without reaction (*dashed line*).

Table E8.7.1: *Transfer enhancement ratio for a range of liquid film coefficients common on the ocean surface*

Wind velocity (m/s):	0.5	1	2	4	6	8	12	16
$\sqrt{kD}/K_L$:	19.7	6.3	2.0	0.7	0.3	0.2	0.1	0.07
K_E/K_L:	19.9	6.5	2.2	1.2	1.2	1.0	1.0	1.0

The flux rate at $z = 0$ is found by the equation

$$J = -D\frac{\partial C}{\partial z}\Big|_{z=0} \tag{E8.7.3}$$

which gives

$$J = C_0\sqrt{kD}\left(\mathrm{erf}\sqrt{kt} + \frac{e^{-kt}}{\sqrt{\pi kt}}\right) \tag{E8.7.4}$$

From equation (E8.7.4), we can see that an equivalent, bulk, liquid-film coefficient, K_E, would be

$$K_E = \sqrt{kD}\left(\mathrm{erf}\sqrt{kt} + \frac{e^{-kt}}{\sqrt{\pi kt}}\right) \tag{E8.7.5}$$

Of course, the concentration boundary layer does not grow indefinitely, but the penetration or the surface renewal theories give us a means of dealing with that fact. If we are going to use the penetration theory, $t \Rightarrow t_e$. If we are going to use the surface renewal theory, $t \Rightarrow 1/\pi r$. With the penetration theory, the transfer enhancement ratio, K_E/K_L, is then given as

$$\frac{K_E}{K_L} = \sqrt{\pi kt_e}\left(\mathrm{erf}\sqrt{kt_e} + \frac{e^{-kt_e}}{\sqrt{\pi kt_e}}\right) \tag{E8.7.6}$$

A similar substitution will give the transfer enhancement ratio for the surface renewal theory.

Now, we need to estimate t_e for the ocean, influenced by wind. We will use the empirical equation developed by Wanninkhof (1992):

$$K_L(\text{m/s}) = 1.25 \times 10^{-6}\ W\ (\text{m/s})^{1.64} \tag{E8.6.7}$$

where W is wind velocity at 10 m height. Using equations (E8.6.7) and (8.33), $t_e = D/(\pi K_L^2)$, the transfer enhancement ratios given in Table E8.7.1 were computed. The diffusivity of carbon dioxide is $\sim 1.7 \times 10^{-9}\ \text{m}^2/\text{s}$, using the predictive techniques in Chapter 3. A significant enhancement of carbon dioxide transfer is apparent at lower wind velocities. The enhancement is minor at wind velocities above 8 m/s, however.

Note that Example 8.7 could have been used for the enhanced transfer of any compound given in Table 8.3. The enhancement ratio is correlated against the ratio, $(kD)^{1/2}/K_L$ in Figure 8.19.

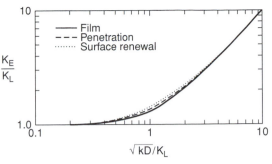

Figure 8.19. The transfer enhancement ratio versus the ratio of $(kD)^{1/2}/K_L$ (Sherwood et al., 1975).

I. Problems

1. (T/F) If a compound is volatile, it will migrate predominately to the air phase.

2. The resistance to mass transfer of a nonvolatile compound is largest in the _____ phase.

3. (T/F) A large Henry's law constant means that the gas is nonvolatile.

4. (T/F) The gas film coefficient is linearly dependent on diffusion coefficient of the compound.

5. Does it make a difference if we use g/m^3 or moles/m^3 for Henry's law constant? Why or why not?

6. How would the distribution of mass differ (qualitatively) if DDT were accidentally spilled into the fish tank of Example 8.4 instead of benzene?

7. You have placed an air diffuser (sparger) at the bottom of a 20-m-deep lake for the purpose of increasing oxygen concentrations. The primary cost of spargers is the cost of pressurizing the air to overcome hydrostatic pressure and head losses. The bubbling action adds oxygen to the water through air bubble–water transfer and also through surface air–water transfer. A steady-state water concentration is reached. Is this steady concentration at equilibrium? Why?

8. Consider Example 8.5, except with (a) toluene (a volatile compound) and (b) trichloromethane (a semivolatile compound) added to the water, instead of toxaphene. Sketch the concentration profiles that would be expected at some point in time.

9. What are the boundary conditions in the environment to which the stagnant film, penetration, surface renewal theories, and laminar boundary layer analogy should be applied? Briefly explain why.

10. Sketch a boundary layer profile for momentum, heat, and mass that would occur in the air and in the water. Explain any differences.

11. A rapid reaction in one phase can increase the overall mass transfer rate. Discuss the influence that bubble size might have on the relative importance of a liquid-phase, rapid reaction on mass transfer rate.

12. The evaporation of water and transfer of SF_6 (a volatile gas tracer) across an air–water interface is measured for three systems:

System	Water evaporative flux (g/m²-s)	SF_6 flux (g/m²-s)
Wind-influenced lake Surface	0.2	10^{-14}
Surface aerator	3.0	10^{-13}
Well-sheltered stream	2×10^{-3}	2×10^{-15}

For all three systems, the following apply:

SF_6 concentration in water $= 10^{-10}$ g/m³
Air temperature $= 20°C$
Dew point temperature $= 5°C$
Air pressure $= 1$ atm

Recognizing that $K_{GA} \sim D_{aA}$ and $K_{LA} \sim D_{wA}$, determine the mass flux rate for the following compounds if their water concentration is 10^{-9} g/m³ and their air concentration is zero: trichloromethane, tribromomethane, benzene, and 1-butanol. Required chemical data is given below.

Chemical	D_{aA} (m²/s)	D_{wA} (m²/s)	H_A (dimensionless concentration units)
Water (H_2O)	2.6×10^{-5}	NA	3×10^{-5}
Sulfur hexafluoride (SF_6)	1.9×10^{-5}	6.2×10^{-10}	82
Benzene (C_6H_6)	0.9×10^{-5}	$10. \times 10^{-10}$	0.18
Trichloromethane ($CHCl_3$)	1.8×10^{-5}	5.5×10^{-10}	0.056
Tribromomethane ($CHBr_3$)	1.7×10^{-5}	5.0×10^{-10}	0.0072
1-Butanol (C_4H_9OH)	0.9×10^{-5}	7.7×10^{-10}	0.00023

13. Round Lake in southwestern Minnesota is down wind from an oil paint factory that has air emissions of solvents. Two of the primary solvents emitted are 1,2-dichloroethylene and heptane, with mean air concentrations over the lake of 2 ppb(v) and 100 ppb(v), respectively. Given the chemical data below, estimate the eventual equilibrium concentration and the mass of these two compounds in the water, suspended particulate, sediments, and the fish of Round Lake. The volume of the lake is 5×10^7 m^3, surface area is 10^7 m^2, particulate concentration is 30 g/m^3 with an organic carbon content of 10%, and the surface sediments have a 2% organic carbon content, with a density of 1.5 g/cc. The organic sediments occupy the upper 0.2 m of the sediment and at deeper depths they are an inorganic glacial till. The fish and other living organic matter add up to 0.03 g/m^3. Assume that the decomposition rates of both compounds can be ignored.

	$C_2H_2Cl_2$	C_7H_{16}
$\log k_{ow}$	2.1	4.66
H_A	0.0067 atm – m^3/mole	2.06 atm – m^3/mole
M_A	97 g/mole	100 g/mole

9 Air–Water Mass Transfer in the Field

The principles of air–water mass transfer are often difficult to apply in field measurements and thus also in field predictions. The reasons are that the environment is generally large, and the boundary conditions are not well established. In addition, field measurements cannot be controlled as well as laboratory measurements, are much more expensive, and often are not repeatable.

Table 9.1 lists some of the theoretical relationships from Chapter 8, for example, and the difficulties in applying these relationships to field situations. Eventually, application to the field comes down to a creative use of laboratory and field measurements, with a good understanding of the results that theory has given us and to make sure that we do not violate some of the basic principles of the theoretical relationships.

The value of β has not been attempted in the field to date. So, how do we determine K_L for field applications? The determination of dynamic roughness, z_0, has also been difficult for water surfaces. The primary method to measure K_L and K_G is to disturb the equilibrium of a chemical and measure the concentration as it returns toward either equilibrium or a steady state. Variations on this theme will be the topic of this chapter.

A. Gas Transfer in Rivers

1. Measurement of Gas Transfer Coefficient

Rivers are generally considered as a plug flow reactor with dispersion. Determination of the dispersion coefficient for rivers was covered in Chapter 6, and determination of the gas transfer coefficient is a slight addition to that process. We will be measuring the concentration of two tracers: a volatile tracer that is generally a gas (termed a gas tracer, C) and a conservative tracer of concentration (C_c). The transported quantity

Table 9.1: *Theoretical relationships for gas transfer coefficient and the difficulties in applying them in the field*

Theoretical relationship	Difficulties in field application
$J = -D\dfrac{\partial C}{\partial z}\Big|_{z \to 0}$	Cannot measure $\partial C/\partial z$ within 100 μm of the free surface
$J = K_L \left(\dfrac{C_g}{H} - C_w \right)$	Must determine K_L
$J = K_G (C_g - HC_w)$	Must determine K_G
$K_G = \dfrac{\kappa^2 \left[u(z) - u(z=0) \right]}{\left[\ln\left(\dfrac{z}{z_{0A}}\right) - \psi_A \right]\left[\ln\left(\dfrac{z}{z_{0m}}\right) - \psi_m \right]}$	Do not know z_{0m} or z_{0A}
$K_L = D/\delta$	Do not know $\partial(t, D)$
$K_L = \sqrt{Dr}$	Do not know r
$K_L \approx \sqrt{D\beta_{\text{rms}}}$	Must determine β_{rms}

will be the ratio of the two tracers, $R = C/C_c$. Then, the convection-dispersion equation becomes

$$\frac{\partial R}{\partial t} + \frac{Q}{A_{cs}}\frac{\partial R}{\partial x} = D_L \frac{\partial^2 R}{\partial x^2} + K_L a \, (R_s - R) \tag{9.1}$$

where Q is the river discharge, A_{cs} is the cross-sectional area, $K_L a = K_L A/V = K_2$ is the reaeration coefficient, and a is the specific surface area or surface area per volume of water. If we are oriented in Lagrangian coordinates, the convection term does not appear. In addition, the gradient of the ratio R is presumed to be low, because as the conservative tracer spreads, so will the gas tracer. Finally, most gas tracers do not have a significant concentration in the atmosphere, so $R_s = 0$. Then, equation (9.1) becomes

$$\frac{\partial R}{\partial t} = -K_2 R \tag{9.2}$$

with solution

$$\ln \left(\frac{R_2}{R_1} \right) = -K_2(t_2 - t_1) \tag{9.3}$$

where subscripts 1 and 2 correspond to times 1 and 2. Most gas transfer measurements are made with a dual tracer pulse, in which the pulse requires some distance to mix across the river, as in Chapter 6. Thus, a tracer cloud is followed, and the time is generally taken to be the center of mass of the cloud on which multiple measurements are made. Then, equation (9.4) becomes that which is actually used in practice:

$$K_2 = -\frac{1}{(t_2 - t_1)} \ln \left(\frac{\sum\limits_{n} R_{2i}}{\sum\limits_{n} R_{1i}} \right) \tag{9.4}$$

where n is the total number of measurements at a given location. There are other considerations with regard to gas transfer measurements in rivers that are detailed

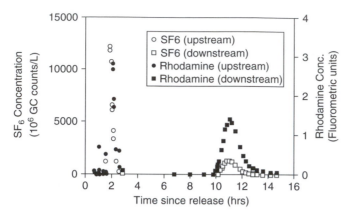

Figure 9.1. Gas tracer pulses for the James River (North Dakota) used to measure the reaeration coefficient. GC, gas chromatograph; SF$_6$, sulfer hexafloride.

elsewhere (Hibbs et al., 1998; Kilpatrick, 1979). Some typical gas tracer pulses are given in Figure 9.1. These measurements are a significant effort, because one needs to be out for 2 or 3 days and mobilize when the tracer cloud passes, which is usually in the middle of the night.

Of course, the typical objective is the K_2 value for oxygen to be used in such things as total daily maximum load calculations, and we just have the K_2 value for the gas tracer. We can use equation (8.62) to get us from the transfer of one compound to another, because β_{rms} is similar for all compounds:

$$\frac{K_2(O_2)}{K_2(\text{tracer})} = \left(\frac{D_{O_2}}{D_t}\right)^{1/2} \tag{9.5}$$

where D_t is the diffusion coefficient of the gas tracer.

2. Prediction of the Gas Transfer Coefficient in Rivers

Measurements of reaeration coefficients have been made at a number of locations over the years, and it is natural that individuals would try to correlate these measurements with the measured parameters of the river so that predictions can be made elsewhere. A partial compilation of measurements is given in Figure 9.2. Although there is scatter in flume measurements, this is exceeded by a factor of 10 in field measurements.

Moog and Jirka (1998) investigated the correspondence of a number of equations with the available data, using the mean multiplicative error, MME:

$$\text{MME} = \exp\left[\frac{\sum_{i=1}^{N}\left|\ln\left(K_p/K_m\right)_i\right|}{N}\right] \tag{9.6}$$

where K_p and K_m are the predicted and measured K_2 values, and N is the number of data. The MME compares the ratio of the prediction with the measurement, so that the study would not be biased toward the higher values of K_2. There are orders

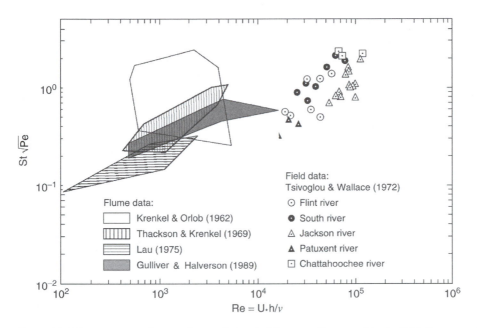

Figure 9.2. Measurements of reaeration coefficient in laboratory flumes and in the field. Present authors are Gulliver and Halverson (1989). St is a Stanton number, K_L/u_*.

of magnitude differences in reaeration coefficients, and to simply take the difference between predictions and measurements as the residuals causes the residuals of the larger reaeration coefficients to dominate the process. The MME thus has the following advantages:

1. The small and large values of K_2 are considered equally.
2. The MME is the geometric mean of K_p/K_m. Thus, a given equation is on average, in error by a factor equal to the MME.
3. The MME of K_2 is equal to the MME of K_L. This occurs because the absolute values of the K_p/K_m ratios were utilized, instead of the square of the ratio.

Moog and Jirka then calibrated the lead coefficient and studied the predictive capability of 10 "calibrated" empirical equations to predict reaeration coefficients that were the result of 331 field studies. The result was surprising, because the best predictive equation was developed by Thackston and Krenkel's (1969) from laboratory flume studies, and the comparison was with field equations. In dimensionless form, Thackston and Krenkel's (1969) calibrated (multiplying the lead coefficient by 0.69) equation can be converted to a dimensionless form utilizing Sherwood, Schmidt, Reynolds, and Froude numbers:

$$Sh = 4.4 \times 10^{-3} Sc^{1/2} Re_* \left(1 + F^{1/2}\right) \tag{9.7}$$

where $Sh = K_L h/D = K_2 h^2/D$, $Sc = \nu/D$, $Re_* = u_* h/\nu$, $F = U/\sqrt{gh}$, h is mean stream depth, D is diffusion coefficient of the gas in the water, ν is kinematic viscosity, $u_* = (ghS)^{1/2}$ is shear velocity, S is water surface slope, U is mean stream velocity, and g is the acceleration of gravity. To convert Thackston and Krenkel's equation to equation (9.7), we have assumed that K_L is proportional to $D^{1/2}$, included the viscosity of water at 20°C, and assumed that $Sc = 476$ for oxygen at 20°C. Equation (9.7) should be functional at all water temperatures in which water is a liquid and for all gases. It is probable that the Froude number in equation (9.7) partially accounts for the additional water surface roughness that occurs above a Froude number of 1.0.

Equation (9.7) results in the following relation for surface renewal rate, r:

$$r = 1.94 \times 10^{-5} \frac{u_*^2}{v} \left(1 + F^{1/2}\right) \tag{9.8}$$

Moog and Jirka also found that equation (9.7), even though it was the best predictor, still had an MME of 1.8. This means that one can expect the predictions of equation (9.7) to be off the field measurements by either multiplying or dividing by a factor of 1.8. Fifty percent of the predictions will differ by more than this factor and 50% by less. In addition, they found that below a stream slope of 0.0004, it is just as good to simply use a constant value of K_2 at 20°C of 1.8 days^{-1}, with a 95% confidence interval corresponding to a multiplicative factor of 8. At the low slope values, other factors, such as wind velocity and surfactants on the water surface, could become influencing factors. Considering that there are 331 "high-quality" measurements of reaeration coefficient in streams, with 54 at $S < 0.0004$, we have not made great advances in predicting reaeration coefficient. The obvious question is Why?

There are four possible reasons that will be presented herein. They will be indicative of the difficulties that exist when taking detailed results of experiments and analysis from the laboratory to the field.

1. *Field measurements are not as precise as laboratory measurements.* Although this is a true statement, some dedicated field experimentalists have improved the field techniques greatly over recent decades (Clarke et al., 1994; Hibbs et al., 1998; Kilpatrick et al., 1979; Tsivoglou and Wallace, 1972). Whereas the implementation of field studies is still a challenge, the accuracy cannot account for MME of 1.8.

2. *Turbulence is generated at the channel bottom, and reaeration occurs at the top of the channel.* The importance of turbulence to gas transfer was illustrated in Sections 8.C and 8.D. One confounding experimental problem is that the turbulence is generated at the bottom of the channel and goes through changes in intensity and scale as it moves toward the water surface. This problem exists in both flume and field channel studies, however, and the flume studies are significantly more precise than the field studies.

3. *The independent parameters that are measured are not truly indicative of the important processes for gas transfer.* In Section 8.D, we discussed the revelations of Hanratty and coworkers – that the process important to gas transfer is two-dimensional divergence on the free surface (Hanratty's β):

$$\beta = -\left(\frac{\partial u}{\partial x} + \frac{\partial v}{\partial y}\right) \tag{8.67}$$

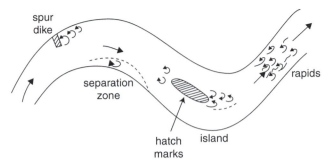

Figure 9.3. Illustration of additional surface vortices in a natural river.

Tamburrino and Gulliver (2002) and Tamburrino and Aravena (2000) have experimentally visualized β on the free surface of turbulent flows and indicated that
there is a correlation between β and vorticity, ω, on the free surface:

$$\omega = \frac{\partial u}{\partial y} - \frac{\partial v}{\partial x} \qquad (9.9)$$

Now, consider a natural river, illustrated in Figure 9.3. There are many sources
of vorticity in a natural river that are not related to bottom shear. Free-surface
vortices are formed in front of and behind islands and at channel contractions and
expansions. These could have a direct influence on reaeration coefficient, without
the dampening effect of stream depth. The measurement of β and surface vorticity
in a field stream remains a challenge that has not been adequately addressed.

4. *The mean values that are determined with field measurements are not appropriate.*
Most predictive equations for reaeration coefficient use an arithmetic mean velocity, depth, and slope over the entire reach of the measurement (Moog and Jirka,
1998). The process of measuring reaeration coefficient dictates that these reaches
be long to insure the accuracy of K_2. Flume measurements, however, have generally shown that $K_2 \sim u* /h$ or $K_2 \sim (S/h)^{1/2}$ (Thackston and Krenkel, 1969;
Gulliver and Halverson, 1989). If this is truly the case, we should be taking the
mean of $S^{1/2}$ and the mean of $h^{-1/2}$ to use the predictive equations to estimate
reaeration coefficient. Example 9.1 will investigate whether this is an important
consideration.

EXAMPLE 9.1: *Estimation of K_2 using measured independent parameters and two types
of mean values*

Most natural rivers and streams are a series of pools and riffles. Calculate the K_2
for one pool riffle pair of equal lengths in a river carrying 12 m³/s at 20°C, using
arithmetic means and the mean values weighted according to equation 9.7. The pool
has a width of 60 m, a mean depth of 2 m, and a bottom roughness of 2 mm. The
riffle has a width of 10 m, a mean depth of 0.5 m, and is a gravel bottom with 20 mm
roughness. No stream slope can be measured over such a short reach.

Equation (9.7) gives us

$$K_2 = 4.4 \times 10^{-3} Sc^{-1/2}(1 + F^{1/2})\frac{u_*}{h} \qquad \text{(E9.1.1)}$$

Because we do not have the slope to determine u_*, we will use the following equation that results from the definition of the Darcy–Weisbach friction factor:

$$u_* = \sqrt{g R_h S} = \sqrt{\frac{f}{8}} U \qquad \text{(E9.1.2)}$$

where U is cross-sectional mean velocity and f is the Darcy–Weisbach friction factor. Moody's diagram, given in Appendix A–1, can be used to determine f. We need the Reynolds number, $Re = 4U R_h/\nu$ and the relative roughness, $\varepsilon/4 R_h$, where $R_h = A/P$ is the hydraulic radius, A is the cross-sectional area, and P is the wetted perimeter. In general, the term $4 R_h$ takes the place of the diameter in these calculations for noncircular cross sections, because the diameter is equal to $4 R_h$ in a circular cross section. We will also calculate the mean stream slope, $S = f U^2/(8g R_h)$, rather than the mean shear velocity, because that is what is measured in practice. Then, equation (E9.1.1) becomes

$$K_2 = 4.4 \times 10^{-3} Sc^{-1/2}(1 + F^{1/2})\frac{(g R_h S)^{1/2}}{h} \cong 4.4 \times 10^{-3} Sc^{-1/2}(1 + F^{1/2})\left(\frac{gS}{h}\right)^{1/2}$$

$$\text{(E9.1.3)}$$

and we can calculate the following parameters for the pool and riffle:

Pool	Riffle
$U = Q/A = 0.1\,\text{m/s}$	$U = 2.4\,\text{m/s}$
$P = 60 + 4 = 64\,\text{m}$, assuming a rectangular cross section	$P = 11\text{m}$
$R_h = A/P = 1.88\,\text{m}$	$R_h = 0.46\,\text{m}$
$Re = 7.5 \times 10^5$	$Re = 4.4 \times 10^6$
$\varepsilon/(4R_h) = 2.7 \times 10^{-4}$	$\varepsilon/(4R_h) = 0.010$
$f \sim 0.0155$	$f \sim 0.038$
$S = 0.0155{*}0.1^2/(8{*}9.8{*}1.88) = 1.1 \times 10^{-6}$	$S = 0.038{*}2.4^2/(8{*}9.8{*}.46)$
	$= 6 \times 10^{-3}$
$F = 0.1/(9.8{*}1.88)^{1/2} = 0.023$	$F = 2.4/(9.8{*}.46)^{1/2} = 1.13$
$Sc = 1 \times 10^{-6}/2.1 \times 10^{-9} = 476$	$Sc = 476$

We can now use these values to calculate the means and K_2:

Arithmetic mean:

$$S = (1.1 \times 10^{-6} + 6 \times 10^{-3})^{1/2} = 3 \times 10^{-3}$$
$$h = (2 + 0.46)/2 = 1.23\text{m}$$
$$F = (0.1 + 2.4)/2/(9.8 * 1.23)^{1/2} = 0.36$$

Then, equation (E9.1.3) gives

$$K_2 = 9.3 \times 10^{-5} \text{ s}^{-1}$$

Means weighted according to equation (E9.1.3)

$$S^{1/2} = [(1.1 \times 10^{-6})^{1/2} + (6 \times 10^{-3})^{1/2}]/2 = 3.9 \times 10^{-2}$$
$$S = 1.5 \times 10^{-3}$$
$$1/h^{1/2} = (1/2^{1/2} + 1/0.5^{1/2})/2 = 1.06 \text{ m}^{-1/2}$$
$$h = 0.89 \text{ m}$$
$$F^{1/2} = (0.023^{1/2} + 1.13^{1/2})/2 = 0.61$$
$$F = 0.37$$

and equation (E9.1.3) gives

$$K_2 = 4.2 \times 10^{-5} \text{ s}^{-1}$$

The difference between the two means is a factor of 2.2. This value is larger than the expected error of equation (9.7). Thus, a channel with a variation in slope and cross section along its length will have a higher K_2 value computed from arithmetic means than an otherwise equivalent channel that does not have variation in slope and cross section. It may not be a coincidence that Moog and Jirka's "calibration" of Thackston and Krenkel's equation for flumes is an adjustment by a factor of 0.69 to represent field measurements. We need to pay attention to the impact that these variations in natural rivers and streams have on our predictive equations for K_2.

The ramifications of the poor K_2 predictive ability are that we cannot do an adequate job of planning for oxygen concentrations during low flow events or for spills, unless we have performed field measurements of reaeration coefficient. This will be explored in Example 9.2.

EXAMPLE 9.2: *Use of predictive relations to determine river withdrawals after a spill*

The Maipo River in central Chile runs past a metal finishing plant in the town of Puente Alto, where 100 kg of carbon tetrachloride (CCl_4) are accidentally spilled into the river. Other compounds were also included in the spill, but the environmental health officials are most concerned about the CCl_4. They have decided that all water supply plants on the river should be shut down if the concentration of the pulse is greater than 5 ppb. At what time and how far downstream will this occur?

In this reach, on the day of the spill, the Maipo River carried a discharge of 60 m³/s, at a mean depth of 2 m, with a mean cross-sectional area of 60 m² and a stream slope of 0.002. The longitudinal dispersion coefficient is known to be 20 m²/s.

The mass transport equation for this problem is as follows:

$$\frac{\partial C}{\partial t} + \frac{Q}{A_{cs}} \frac{\partial C}{\partial s} = D_L \frac{\partial^2 C}{\partial s^2} - K_2 C \qquad \text{(E9.2.1)}$$

where C is the concentration of CCl_4, Q is river discharge, A_{cs} is cross-sectional area, D_L is the longitudinal dispersion coefficient, and s is the coordinate along the river centerline. Both D_L and K_2 are critical to the solution of this problem, and both have a significant uncertainty in prediction. For this application, equation (E9.2.1) has the following boundary conditions:

1. At $x = 0$ and $t = 0$, a total mass of $M = 100$ kg CCl_4 is released as a pulse into the river.
2. At $t \Rightarrow$ large, and $x \Rightarrow$ large, $C \Rightarrow 0$.

From analogy with Example 2.8, adding in a convection term, equation (E9.2.1) with these boundary conditions has solution

$$C = \frac{M/A_{cs}}{\sqrt{4\pi D_L t}} \exp\left[-\frac{(s - Ut)^2}{4D_L t} - K_2 t \right] \qquad (E9.2.2)$$

The peak of equation (E9.2.2) occurs at $s = Ut$, so the peak concentration is given by

$$C_{\text{peak}} = \frac{M/A_{cs}}{\sqrt{4\pi D_L t}} e^{-K_2 t} \qquad (E9.2.3)$$

The reaeration coefficient will indicate the transfer of CCl_4, because it is a volatile compound, as long as we use the correct Schmidt number in equation (9.7). At $20°C$, CCl_4 has a diffusion coefficient of 1.2×10^{-9} m^2/s. Then, $Sc = 833$, and equation (9.7) gives $K_2 = 2.2 \times 10^{-5}$ s^{-1}.

As described in equation (6.59), longitudinal dispersion coefficient has a 67% confidence interval that is a factor of 1.7 times the best estimate. If the distribution of multiplicative uncertainty is normal, the 95% confidence interval would be at a factor of 3.4 times the best estimate. The reaeration coefficient has are MME of 1.8 for the Thackston and Krenkel equation (equation (9.7)). Again, if the multiplicative distribution is normal, the MME is 0.4 times the 95% confidence interval. Then the 95% confidence interval is a multiplicative factor of 4.5.

We will use the means and 95% confidence intervals for both D_L and K_2 to determine the time, and distance through $s = Ut$, when $C_{\text{peak}} = 0.0005$ g/m^3. These conditions are listed in Table E9.2.1, with the times determined through iteration on equation (E9.2.3). Table E9.2.1 shows that the peak value of concentration is no longer sensitive to longitudinal dispersion coefficient after roughly 3 days, because the peak is widely spread. The time when the water treatment plants downstream of the spill could turn on the water intake, however, would likely be sensitive to longitudinal dispersion coefficient.

The time and distance before the peak concentration of CCl_4 is below 0.5 ppb is sensitive to uncertainty in the value of K_2. We can see that, within a 95% confidence interval, the time when water intakes need to be turned off could vary between 0.8 and 12.4 days, or over 67 to 1,071 km of the Maipo River. The Maipo River, however, is not 1,071 km long, so the entire river below Puente Alto would need to be alerted.

Table E9.2.1: *Expected variation in the times and distance that water supply needs to be shut down on the Maipo River below the Puente Alto spill to maintain CCl₄ concentrations below 5 ppb*

D_L (m²/s)	K_2 (/s)	t (s)	s (km)	t (hrs)	t (days)	Comments
20	2.2×10^{-5}	269,074	269	75	3.1	Best estimate
68	2.2×10^{-5}	243,829	244	68	2.8	$D_L \times 3.4$, upper CI
5.9	2.2×10^{-5}	294,435	294	82	3.4	$D_L/3.4$, lower CI
20	1.0×10^{-4}	66,703	67	19	0.8	$K_2 \times 4.5$, upper CI
20	5.0×10^{-6}	1,071,219	1,071	298	12.4	$K_2/4.5$, lower CI

CI, confidence interval.

B. Gas Transfer in Lakes, Estuaries, and Oceans

The influence of wind is predominant in determining the liquid film coefficient for lakes, reservoirs, oceans, and many estuaries. Wind creates a shear on the water surface and generates turbulence below and on the water surface. Thus, this section deals with the measurement and prediction of the wind influence on liquid film coefficient.

1. Opportunistic Measurement of Wind Influence

The opportunistic measurement techniques generally used are ^{14}C absorption and ^{222}Rn disequilibrium (Asher and Wanninkhof, 1998). First, there is an estimate of a long-term ($\sim$1,000 years) global gas transfer coefficient of $K_L = 6 \times 10^{-5}$ m/s, developed by assuming steady state between pre-1950 ^{14}C radioactive decay in the oceans and absorption from the atmosphere (Broecker and Peng, 1982). In addition, nuclear testing since 1950 has increased ^{14}C concentration in the atmosphere. Thanks to the atomic testing "battle" between the United States of America and the Soviet Union, we currently have a tracer that can be used on an oceanwide basis. A box model of an ocean basin is still needed. By using an appropriate oceanic model to estimate the depth of the interactive layer, and taking sufficient measurements of ^{14}CO$_2$ at the ocean boundaries and inside the control volume, the fluxes and mean control concentration, respectively, can be determined. Then, the remainder of the flux is assigned to atmospheric fluxes of ^{14}CO$_2$, and a liquid film coefficient is determined from a mass conservation equation:

$$V\frac{\partial\,^{14}C_{\text{basin}}}{\partial t} = \text{Ocean flux rate } (^{14}C_{\text{in}} - {}^{14}C_{\text{out}}) + K_L A(^{14}C_{\text{atm}} - {}^{14}C_{\text{basin}}) \quad (9.10)$$

where V is the volume of the interactive layer and A is the surface area of the ocean basin. An important consideration for these estimates is the depth of the interactive layer (Duffy and Caldera, 1995). The response time of equation (9.10) is on the order of years to decades, so any relationship to wind velocity is a long-term average.

By contrast, the gas transfer estimates utilizing [222]Rn measurements assumes steady state between [222]Rn production from radioactive decay of nonvolatile [226]Rd and gas transfer with the atmosphere. This assumption is possible because [222]Rn has a half-life of only 3.8 days, so accumulation and lateral ocean fluxes of [222]Rn is assumed to be minimal. Again, a potential problem is the active, versus inactive layer of the ocean; in this case, the mixed layer depth that may change during an experiment.

The results of both ^{14}C and [222]Rn measurements of liquid film coefficient versus wind velocity are plotted in Figure 9.4, along with two parameterizations that will be discussed in prediction of wind influence.

2. Measurement of Wind Influence with Deliberate Tracers

Batch Technique. As with river reaeration measurements, tracers can also be put into lakes, estuaries, and oceans to measure the influence of wind on liquid film coefficient. If we have a volatile tracer in a lake with a well-established mixed layer, for example, we can apply the same batch reactor equation from Section 6.A, as though we had a well-mixed tank:

$$V\frac{\partial C}{\partial t} = -K_L A\left(\frac{C_g}{H} - C\right) \tag{9.11}$$

where V is the volume of the mixed layer and A is the interfacial area. If we assume that C_g is a constant value, then we can assign $C' = Cg/H - C$, separate variables, and integrate to achieve

$$\ln\left(\frac{C - C_g/H}{C_0 - C_g/H}\right) = -K_L\frac{A}{V}(t - t_0) = -K_L a(t - t_0) \tag{9.12}$$

where C_0 is the concentration at $t = t_0$ and "a" is the specific surface area, which has units of length^{-1}. A plot of the log term in equation (9.12) versus time will result in a straight line of slope $K_L a$. If there are more than 11 points, the standard error of the slope is approximately the precision (random) uncertainty to the 67% confidence interval. That means that 67% of the data, if Gaussian around the mean, will fall within the confidence interval. Bias (systematic) uncertainty, of course, needs to be analyzed separately. There are other means of determining $K_L a$ for a batch reactor, the best known being the American Society of Civil Engineers (1992) technique that uses a nonlinear regression to determine C_g/H and $K_L a$.

The batch reactor analysis is a natural one to use for laboratory tanks and wind–wave facilities. In addition, it has been used for lakes, which are either well-mixed vertically or where the surface mixed layer is at close to a constant thickness (Livingstone and Imboden, 1993; Torgersen et al., 1982; Upstill-Goddard et al., 1990; Wanninkhof et al., 1987). Typically, sulfur hexafluoride (SF$_6$) or ^{3}He are used as deliberate gas tracers input to the lakes because they are detectable at low concentrations. The results of selected measurements are plotted versus the wind speed Reynolds number in Figure 9.5. The primary difference between the two sets of measurements is the fetch length of the water body, where the Upstill-Goddard et al. measurements were taken on a smaller water body. In general, the uncertainty of one given measurement

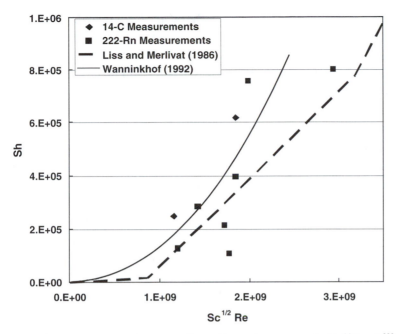

Figure 9.4. Sherwood number versus Reynolds number measured with ^{14}C and ^{222}Rn tracers. *Solid* and *dashed lines* are empirical relations developed from measurements.

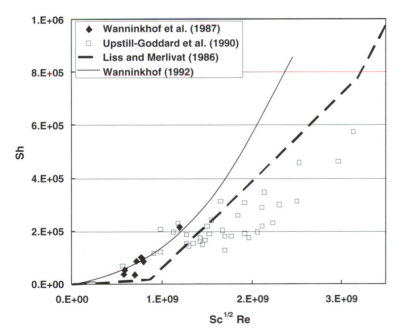

Figure 9.5. Sherwood number versus $Sc^{1/2}\,Re$, measured by the batch reactor technique in small lakes. *Lines* represent equations developed from field measurements.

is large, but a number of measurements, taken together while leaving in the ones that give a negative K_L, results in the summary field measurements seen in Figure 9.5. The problem is that the investigator needs to be involved in deciding which measurements to average, so there is a potential bias in the results. Changes in mixed layer depth need to be taken into consideration, as well, if one is not able to assume a well-mixed lake.

Dual Tracer Technique. The dual tracer measurement technique utilizes two gas tracers with diffusion coefficients that are substantially different, such as ^{3}He and SF_6. This technique can also be utilized with one volatile (gas) tracer and one nonvolatile tracer. We will derive the relevant equations to determine liquid film coefficient from the diffusion equation for both cases, beginning with the two gas tracers.

Consider a cylindrical control volume of depth h that moves with the mean velocity of the tracer cloud containing two gas tracers, designated A and B. Using the cylinder as our control volume, the transport relation for each of the gas tracers can be written as

$$\frac{\partial \overline{C_i}}{\partial t} = D_x \frac{\partial^2 \overline{C_i}}{\partial x^2} + D_y \frac{\partial^2 \overline{C_i}}{\partial y^2} - \frac{K_L}{h} C_{si} \qquad (9.13)$$

where $\overline{C_i}$ is the mean concentration of compound i over the column depth, D_y and D_x are the dispersion coefficients in the horizontal directions, and C_{si} is the concentration of compound i at the water surface. We will perform our mass balance on the ratio, $R = C_A/C_B$. Then, the derivative of R with respect to time is given as

$$\frac{\partial R}{\partial t} = \frac{\partial}{\partial t}\left(\frac{C_A}{C_B}\right) = \frac{1}{C_B}\left(\frac{\partial C_A}{\partial t} - R\frac{\partial C_B}{\partial t}\right) \qquad (9.14)$$

Assuming that the dispersion of the ratio of concentrations to be small, we will ignore the dispersion of R, and equations (9.13) and (9.14) are combined to give

$$\frac{\partial R}{\partial t} = -\frac{1}{h}\left(K_{LA}R_{sA} - K_{LB}RR_{sB}\right) \qquad (9.15)$$

where $R_{sA} = C_{sA}/\overline{C_B}$ and $R_{sB} = C_{sB}/\overline{C_B}$. We will assume the 1/2 power relationship between liquid film coefficient and diffusion coefficient and apply a surface renewal type of relationship. Then

$$\frac{K_{LB}}{K_{LA}} = \sqrt{\frac{D_B}{D_A}} \qquad (9.16)$$

Substituting equation (9.16) into (9.15), and rearranging gives us an equation that can be used to develop K_{LA} from measurements of both tracers:

$$K_{LA} = -\frac{h}{\left(R_{sA} - \sqrt{\dfrac{D_B}{D_A}} RR_{sB}\right)}\frac{\partial R}{\partial t} \qquad (9.17)$$

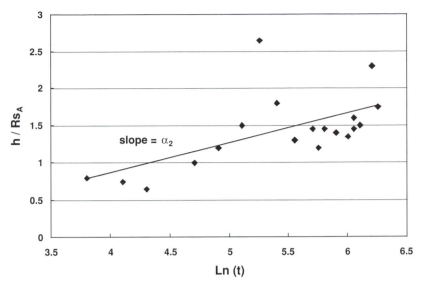

Figure 9.6. Measured data from a dual tracer cloud and the curve fit to the data.

R, R_{sA}, and R_{sB} can all be measured at various times. Then, a plot of the various terms in equation (9.17) will provide us with an estimate of K_L. If tracer B is nonvolatile, $K_{LB} = 0$, and equation (9.17) becomes

$$K_L = -\frac{h}{R_{sA}}\frac{\partial R}{\partial t} \qquad (9.18)$$

Equation (9.17) applies to two volatile tracers and equation (9.18) to one volatile and one nonvolatile tracer.

Gulliver et al. (2002) applied equation (9.18) to field measurements of SF_6 and Rhodamine-WT, a nonvolatile tracer. Their analysis technique was as follows:

1. Regress $\ln(h/R_{sA})$ versus $\ln t$ in a linear regression. Then

$$h/R_{sA} = \alpha_1 t^{\alpha_2} \qquad (9.19)$$

   where α_1 and α_2 are fitted constants. A sample for one data set, shown in Figure 9.6, indicates that there is considerable scatter, which is due to sampling uncertainty (i.e., one does not know if they are at the peak of the tracer cloud).
2. Regress R versus $t^{1-\alpha_2}$. The α_2 power is used to make the dimensions work out properly in the relation for K_L. Then

$$R = \alpha_3 t^{1-\alpha_2} + \alpha_4 \qquad (9.20)$$

A sample of this regression is shown in Figure 9.7, again indicating considerable uncertainty. Equation (9.20) is chosen to result in the proper units for K_L. Combining equations (9.18) to (9.20) yields

$$K_L = \alpha_1 * t^{\alpha_2} * \left[\frac{\partial}{\partial t}\left(\alpha_3 * t^{(1-\alpha_2)} + \alpha_4\right)\right] \qquad (9.21)$$

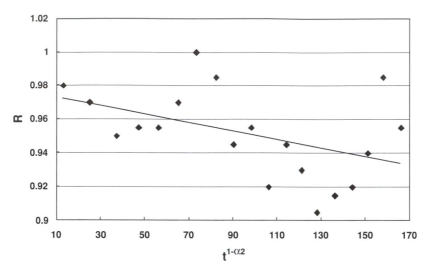

Figure 9.7. Measured data from a dual tracer experiment for 1 day and the curve fit to the data.

or

$$K_L = -\alpha_1 * \alpha_3 * (1 - \alpha_2) \tag{9.22}$$

With the dimensional formulation of equation (9.22), α_1 has units of length/time$^{\alpha 2}$ and α_3 has units of time$^{\alpha 2-1}$. Thus, K_L has units of length/time.

   The precision uncertainty associated with field sampling is generally much larger than that associated with analytical technique, which is roughly $\pm 2\%$ to the 67% confidence interval for the two compounds used as conservative and gas tracers. A technique to determine the precision uncertainty associated with field sampling and incorporated into the mean K_L estimate will therefore be propagated with the first-order, second moment analysis (Abernathy et al., 1985):

$$U_{K_L}^2 = \left(U_{\alpha_1} * \frac{\partial K_L}{\partial \alpha_1} \right)^2 + \left(U_{\alpha_3} * \frac{\partial K_L}{\partial \alpha_3} \right)^2 + \left(U_{\alpha_2} * \frac{\partial K_L}{\partial \alpha_2} \right)^2 \tag{9.23}$$

When the partial derivatives are taken from equation (9.22), equation (9.23) becomes

$$U_{K_L}^2 = [\alpha_3 * (1 - \alpha_2) * U_{\alpha_1}]^2 + [\alpha_1 * (1 - \alpha_2) * U_{\alpha_3}]^2 + (\alpha_1 * \alpha_3 * U_{\alpha_2})^2 \tag{9.24}$$

The variables U_{α_1}, U_{α_2}, and U_{α_3} are the corresponding uncertainty values for each parameter. They are computed to the 67% confidence interval by taking the standard error of each parameter in the regressions (i.e., α_1, α_2 and α_3, and multiplying by their Student t-score t_S (i.e., $U_{\alpha_1} = t_S * SE_{\alpha_1}$), where t_S is the Student t-score at the confidence level of interest and $SE_{\alpha 1}$ is the corresponding standard error for the parameter α_1. The period can be chosen based on the maximum r^2 value or another statistical parameter. The results of four experiments are given in Figure 9.8.

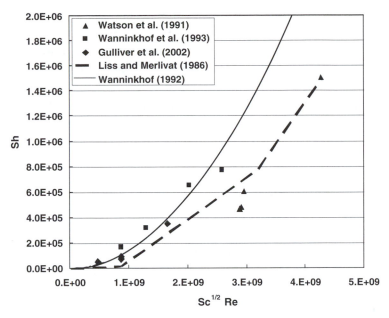

Figure 9.8. Sherwood number measured by the dual tracer technique versus Reynolds number.

Other investigators have used two measurements of the gas tracers taken over a longer time period (several days) and assumed that the cloud of tracers was well-mixed vertically. In this case, the following conditions must be met: (1) changes in the ratio of tracer concentrations caused by dispersion should be negligible compared with those resulting from gas transfer, (2) the water column should be well-mixed vertically, and (3) the experimental area should be close to constant depth. Criteria 1 is satisfied by having a tracer cloud that is large, compared with the depth and by staying close to the center of this cloud. Example 9.2 has shown us that, even at the peak of the concentration cloud, dispersion can have an impact on concentrations. With these assumptions, equation (9.17) becomes

$$\frac{1}{R}\frac{dR}{dt} = \frac{K_{LA}}{h}\left(1 - \sqrt{\frac{D_B}{D_A}}\right) \tag{9.25}$$

which can be integrated to give

$$K_{LA} = \frac{h}{\Delta t}\left(1 - \sqrt{\frac{D_B}{D_A}}\right)(\ln R_{t_2} - \ln R_{t_1}) \tag{9.26}$$

The results of these measurements are also given in Figure 9.8. Unfortunately, the technique utilizing equation (9.26) cannot provide an estimate of sampling uncertainty.

The data contained in Figures 9.4, 9.5, and 9.8 indicate similarity against wind velocity, even though the fetch length and size of the water bodies are different.

This is generally seen as an indication that wind is an important driving factor for lakes, estuaries, and oceans. It has been shown that breaking waves and water surface slicks are important (Asher and Wanninkhof, 1998), and there are other parameters – such as mean square water surface slope – that have been proposed as better indicators (Jähne, 1991). The problem is that our ability to predict these indicators from wind velocity measurements have not been developed and tested for liquid film coefficient.

One other measurement technique that has been used to measure K_L over a shorter time period, and is thus more responsive to changes in wind velocity, is the controlled flux technique (Haußecker et al., 1995). This technique uses radiated energy that is turned into heat within a few microns under the water surface as a proxy tracer. The rate at which this heat diffuses into the water column is related to the liquid film coefficient for heat, and, through the Prandtl–Schmidt number analogy, for mass as well. One problem is that a theory for heat/mass transfer is required, and Danckwert's surface renewal theory may not apply to the low Prandtl numbers of heat transfer (Atmane et al., 2004). The controlled flux technique is close to being viable for short-period field measurements of the liquid film coefficient.

3. Prediction of the Wind-Influenced Gas Transfer Coefficient

There are two predictive relationships based on wind speed. Liss and Merlivat (1986) used a physical rationale to explain the increase in the K_L versus wind speed slope at higher wind velocities in wind–wave tunnel and lake measurements, resulting in a piecewise linear relationship with two breaks in slope. These breaks are presumed to occur at the transition between a smooth surface and a rough surface and between a rough surface and breaking waves. In dimensionless form, this relationship is given as

$$Sh = 3.4 \times 10^{-5} Sc^{1/3} Re \qquad\qquad U_{10} \geq 3.6\,\text{m/s} \qquad\qquad (9.27\text{a})$$

$$Sh = 1.9 \times 10^{-4} Sc^{1/2} (Re - Re_0) \quad 3.6\,\text{m/s} < U_{10} \leq 13\,\text{m/s} \qquad (9.27\text{b})$$

$$Sh = 4.1 \times 10^{-4} Sc^{1/2} (Re - Re_b) \quad 13\,\text{m/s} \leq U_{10} \qquad\qquad (9.27\text{c})$$

where U_{10} is the wind speed at 10 m above the mean water surface, $Sh = K_L L/D$, $Re = U_{10} L/\nu$, $Re_0 = 3.4\,\text{m/s}\, L/\nu$, and $Re_b = 8.3\,\text{m/s}\, L/\nu$. The length scale, L, could be any relevant length scale because it will drop out of the relationship when determining K_L.

Wanninkhof (1992) developed one relation from ^{14}C data:

$$Sh = 3.4 \times 10^{-5} Sc^{1/2} Re\, U_{10}\,(\text{m/s}) \qquad\qquad (9.28)$$

Equation (9.28) does not lend itself to an easy conversion to dimensionless parameters because $K_L \sim U_{10}^2$. It is one equation, however, instead of three, which makes it easier to use in computer programs and spreadsheets. If we assume that we are using $L = 10\,\text{m}$ and $\nu = 1 \times 10^{-6}\,\text{m}^2/\text{s}$, then equation (9.28) becomes

$$Sh = 3.4 \times 10^{-12}\,Sc^{1/2}\,Re^2 \tag{9.29}$$

Both equations (9.27) and (9.29) are compared with the existing field data in Figures 9.4, 9.5, and 9.8. With the scatter in the data, it is difficult to select one equation over another.

Jähne et al. (1984, 1987) proposed that liquid film coefficient is better related to mean water surface slope. Frew (1997) has found that the K_L relationship using mean square slope can be used to describe gas transfer with and without surface slicks. The problem with mean surface slope is that it cannot be accurately predicted for water bodies, because most investigators have emphasized the larger and longer waves, and the slope is most significant for the small, short waves. This will likely be the subject of future investigations.

C. Transfer of Nonvolatile Compounds

The evaporation of water is generally used to determine the gas film coefficient. A loss of heat in the water body can also be related to the gas film coefficient because the process of evaporation requires a significant amount of heat, and heat transfer across the water surface is analogous to evaporation if other sources and sinks of heat are taken into account. Although the techniques of Section 8.D can be used to determine the gas film coefficient over water bodies, they are still iterative, location specific, and dependent on fetch or wind duration. For that reason, investigators have developed empirical relationships to characterize gas film coefficient from field measurements of evaporation or temperature. Then, the air–water transfer of a nonvolatile compound is given as

$$J_A = K_G\,[C_A(z) - C_A(z = 0)] \tag{9.30}$$

where $C_A(z)$ is the concentration of compound A at an elevation z, which is typically the elevation of wind and temperature/humidity measurements to compute K_G.

The relationships developed from field measurements have been made dimensionless with the assumptions that $\nu = 1.33 \times 10^{-5}\,\text{m}^2/\text{s}$ and $D_{\text{H}_2\text{O}} = 2.6 \times 10^{-5}\,\text{m}^2/\text{s}$ to facilitate comparisons between relations and avoid dimensional problems. They are given in Table 9.2. The early measurements were to investigate the loss of water from the reservoirs of the Colorado River in the United States, and the later measurements were designed to investigate heat loss from heated water bodies. A revelation occurred in 1969, when Shulyakovskyi brought in buoyancy forces as related to natural convection to explain the heat loss from heated water at low wind velocities. This was picked up by Ryan and Harleman (1973), who realized that natural convection could explain the need for a constant term in front of the relationship for gas film coefficient, as had been found by Brady et al. (1969), Kohler (1954), Rymsha and Dochenko (1958), and Shulyakovskyi (1969). Finally, Adams et al. (1990) rectified the overprediction of Ryan and Harleman's formulation at high wind velocities with the root sum of squares relation and brought in Harbeck's (1962) relation to include a term that represents fetch dependence.

By using the latent heat of vaporization for water, L_e, from equation (8.86), the previous relations for mass transfer can also be used for heat transfer due to evaporation:

$$\varphi_e = \frac{L_e K_G}{\rho C_P}(T_2 - T_s) \tag{9.31}$$

where ρ is the density of air and C_P is the heat capacity of air at constant pressure. We will compare the gas film coefficients from Table 9.2 in Example 9.3.

EXAMPLE 9.3: *Application of characteristic relations for gas film coefficient*

The WECAN, Inc. consulting company has a project that requires determining the evaporation from the 10 hectare cooling pond at the Hang Dog Power Facility, and they realized that they do not know how to determine the gas film coefficient. A table similar to Table 9.2 was found, and they decided to compare the resulting predictions to see if it made a significant difference. We will duplicate their results for one such comparison under the following conditions:

$$\text{Water temperature, } T_s, = 30°C$$
$$\text{Air temperature, } T = 10°C$$
$$\text{Relative humidity at 2 m height} = 40\%$$
$$\text{Air pressure} = 1 \text{ atm}$$
$$\text{Various wind velocities up to 15 m/s}$$

The *Handbook of Chemistry and Physics*, or some of the material in the Appendix A–3, can be used to determine the following fluid properties:

$\nu = 1.33 \times 10^{-5} \text{ m}^2/\text{s}$
$D = 2.4 \times 10^{-5}$ for water vapor, from Chapter 3
$\alpha = 2.0 \times 10^{-5} \text{ m/s}$

Now, virtual temperature is given by the equation

$$\Delta\theta_v = \| T_s(1 + 0.378 p_{vs}/P_a) - T_a(1 + 0.378 p_{va}/P_a), 0 \| \tag{8.73}$$

and saturation vapor pressure will be required to compute virtual temperature:

$$p_{vs}(mb) = 6.11 \exp\left\{\frac{17.3[T(°K) - 273]}{T(°K) - 35.9}\right\} \tag{8.74}$$

Then, $p_{vs}/P_a = 0.042$ at 30°C and $= 0.012$ at 10°C, and

$$p_{va} = \frac{RH}{100} p_{vs} \tag{8.75}$$

Table 9.2: *Relationships developed to characterize gas film coefficient over water surfaces*

Investigator	Formula	Water body
Adams et al. (1990)	$Sh = \left[\left(0.125\, Ra_V^{1/3}\right)^2 + \left(0.0061\, Sc\, Re^{0.933}\, \hat{W}^{0.033}\right)^2\right]^{1/2}$	East Mesa Geothermal Facility and cooling ponds
Brady et al. (1969)	$Sh = 0.049 \left(\dfrac{Ra}{\beta_T \Delta T}\right)^{1/3} + 9.6 \times 10^{-6}\, Sc\, Re^{1/3}\, \hat{W}^{1/3}$	Power plant cooling ponds
Ficke (1972)	$Sh = 0.0019\, Pe$	Pretty Lakes, Ind
Gulliver and Stefan (1986)	$Sh = 0.125\, Ra_V^{1/3} + 0.0013\, Pe$	Heated streams
Harbeck (1962)	$Sh = 0.0061\, Sc\, Re^{0.933}\, \hat{W}^{0.033}$	Various reservoirs
Harbeck et al. (1958)	$Sh = 0.002\, Pe$	Lake Mead, Ariz
Hughes (1967)	$Sh = 0.0014\, Pe$	Salton Sea, Calif
Kohler (1954)	$Sh = 0.011 \left(\dfrac{Ra}{\beta_T \Delta T}\right)^{1/3} + 7.5 \times 10^{-5}\, Pe$	Lake Hefner, Ariz
Marciano and Harbeck (1954)	$Sh = 0.0017\, Pe$	Lake Hefner, Ariz
Ryan and Harleman (1973)	$Sh = 0.125\, Ra_V^{1/3} + 0.0016\, Pe$	Cooling ponds
Rymsha and Dochenko (1958)	$Sh = 0.042 \left(\dfrac{Ra}{\beta_T \Delta T}\right)^{1/3} + 0.0125\, Ra^{1/3} + 0.0016\, Pe$	Various rivers, heated in winter
Turner (1966)	$Sh = 0.0031\, Pe$	Lake Michie, NC
Shulyakovskyi (1969)	$Sh = 0.031 \left(\dfrac{Ra}{\beta_T \Delta T}\right)^{1/3} + 0.125\, Ra_V^{1/3} + 0.0017\, Pe$	Various water bodies

$$Sh = \frac{K_G A^{1/2}}{D} \qquad Ra = \frac{g\beta_T \Delta T A^{3/2}}{D\nu}$$

$$Pe = \frac{W A^{1/2}}{\nu} \qquad Ra_V = \frac{g\beta_T \Delta T_V A^{3/2}}{D\nu} \qquad \hat{W} = \frac{W^2}{g A^{1/2}}$$

K_G, gas film coefficient; A, surface area of water body; D, diffusion coefficient of compound in air; W, wind velocity at 2 m above the mean water surface; ν, kinematic viscosity of air; α, thermal diffusion coefficient of air; g, acceleration of gravity; β_T, thermal expansion coefficient of moist air; ΔT, temperature difference between water surface and 2 m height; ΔT_V; virtual temperature difference between water surface and 2 m height.

So that $p_{va}/P_a = 0.005$. Then, $\Delta\theta_V = 20.5°C$. Finally,

$$\beta_T = \frac{1}{T_{av}(°C) + 273} \tag{8.76}$$

So that $\beta_T = 3.29 \times 10^{-3}°K^{-1}$. The results for five formulas are given in Figure E9.3.1. The major differences are the slope at high wind speeds and the intercept at a wind velocity of 0. Probably the best documented is the relation of Adams et al. (1990), which used much of the previous field data and tested the results on two additional water bodies at a variety of buoyancy parameters. This relation transitions between the Ryan and Harleman (1973) equation at low wind velocity and that of Harbeck (1962) at a higher wind velocity.

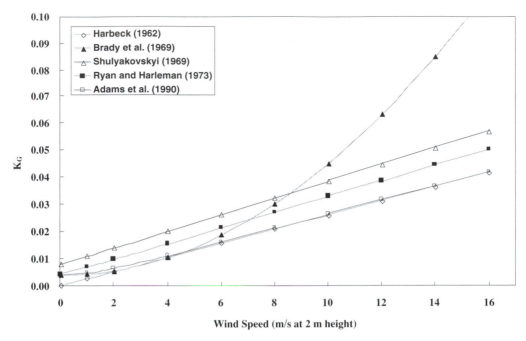

Figure E9.3.1. Comparison of five formulas for gas transfer coefficient versus wind speed for example 9.3.

D. Gas Transfer from Bubbles

Measurement of Liquid Film Coefficient

To determine the liquid film coefficient for gas transfer from or into bubbles, dissolved oxygen and dissolved nitrogen concentrations must be measured (Schierholz et al. 2006) This is possible with a polargraphic probe for dissolved oxygen and a total gas meter, which measures the pressure of gas in equilibrium with the water. Nitrogen (+ argon) concentration is found by subtraction of dissolved oxygen and water vapor from total gas concentration. Then, the equations in Section 8.F may be used to determine the value of $K_L A_b$ and K_{LS} that will give these concentration measurements. Equation (8.102) changes with depth as z and y_m change with depth. The boundary condition of equations (8.101) and (8.102) is the gas molar ratio at the sparger, which is 0.266 for air and the initial concentrations C_O and C_N in the liquid. The equations are then solved from $z = h_d$ to $z = 0$ at one moment in time and used to compute C_O and C_N in equation (8.90) for the next time step. The resulting concentration curves are adjusted by changing $K_{LO} A_b$ and K_{LSO} until the concentration versus time curves match the measured one for C_O. Thus, two transfer coefficients, $K_{LO} A_b$ and K_{LSO}, can be determined.

Schierholz et al. (2006) applied equations (8.98) to (8.102) to various tank experiments of depths that varied from 2.25 to 32 m in depth. The results were that a value of the bubble transfer coefficient, $K_L A_b / V$ and the surface transfer coefficient, K_{LS}, could be determined. The following relationship were fit to the resulting bubble–water transfer coefficients:

$$Sh = \frac{1}{6}\frac{We^{3/5} Sc^{1/2} Re}{Fr} \qquad \text{Fine bubble diffusers} \qquad (9.32)$$

and

$$Sh = \frac{1}{37}\frac{We^{3/5} Sc^{1/2} Re}{Fr} \qquad \text{Course bubble diffusers} \qquad (9.33)$$

where $Sh = K_L A_b h_d^2/DV$, $We = U^2 h_d \rho/\sigma$, $Sc = v/D$, $Re = U h_d/v$, and $Fr = U/\sqrt{g h_d}$ is the Froude number. In the dimensionless numbers, h_d is diffuser depth, V is the volume of the water body, and U is the superficial gas velocity, Q_a/A_{cs}, where Q_a is the air discharge at standard temperature and pressure and A_{cs} is the cross-sectional area of the water body or tank. A fine bubble diffuser has a bubble diameter leaving the diffuser that is less than roughly 4 mm in diameter. The resulting water surface transfer coefficients were fit to the following equation:

$$Sh_s = 49 \, Sc^{1/2} \, Re \left(\frac{A_{cs}}{h_d^2}\right)^{0.72} \qquad (9.34)$$

where Sh_s is the Sherwood number for surface transfer, $K_L A_s/(h_d D)$, and A_s is the surface area of the water body. It is interesting that K_{Ls} is linearly dependent on gas flow rate. This is likely because the bubbles passing through the surface create a significant free-surface turbulence. Of course, these tests were performed without wind or a mean flow, so any wind or low influence would need to be somehow factored in. Figures 9.9 and 9.10 provide the data analysis results and the curvefits of the characteristic equations.

EXAMPLE 9.4: *Sizing bubble diffusers for a reservoir with combined sewer overflow*

The McCook, Thornton, and O'Hare reservoirs make up the Chicago-land Underflow Plan, an integral part of Chicago's $3 billion Tunnel and Reservoir Plan. This system of intercepting sewers, dropshafts, tunnels, and reservoirs will capture and store combined sewage and stormwater until municipal water reclamation plants can treat it. Applying equations (9.31) through (9.33), and equations (8.90), (8.101), and (8.102), both a coarse bubble and a fine bubble aeration system for the McCook reservoir in Chicago can be designed. The aeration design needs to maintain aerobic conditions in the reservoir (at least 2 mg/L) that has a designed area of 395,300 m². The 1-in-100-year event in July should cause the McCook Reservoir to fill to its maximum level of 73 m (Robertson, 2000). When the reservoir is at its maximum elevation after a big flood, the biochemical oxygen demand (BOD) will be lower because of dilution of the sanitary sewage with stormwater. A 5-day BOD design range of 30 mg/L when full to 80 mg/L when at the lowest depth of 10 m is estimated and a BOD decay rate of 0.25 d⁻¹ is assumed. Other design criteria are that the diffusers will be 1 m above the bottom of the reservoir, the maximum air flow rate per

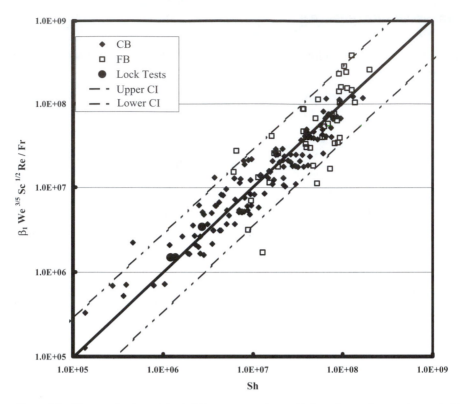

Figure 9.9. Characterization of the bubble mass transfer coefficients for Tank tests. Coarse bubble (CB) $\beta_1 = 1/36$, and fine bubble (FB) $\beta_1 = 1/6$. The 95% confidence interval is included (CI) (Schierholz et al. 2006).

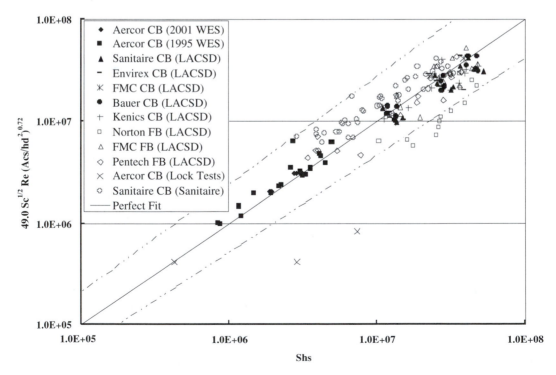

Figure 9.10. Characterization of the surface mass transfer coefficients for the tank tests. The 95% confidence interval is included (Shierholz et al., 2006). CB, coarse bubble; FB, fine bubble; LACSD, Los Angeles County Sanitary District; WES, U.S. Army Engineer Waterways Experiment Station.

coarse bubble diffuser is 100 scmh, and the maximum air flow rate per fine bubble diffuser is 20 scmh.

Using an iterative procedure between equations (8.88) through (8.102) and equations (9.32) through (9.34), an aeration system to accommodate the BOD range can be designed. Using coarse bubble diffusers and the high BOD of 80 mg/L, assumed to occur at a depth of 10 m, it was determined that a single coarse bubble diffuser with an air flow rate of 100 scmh would maintain aerobic conditions for an area of 341 m^2 with the above conditions. Based on this, 1,160 coarse bubble diffusers supplying a total air flow of 116,000 scmh would be needed for the McCook Reservoir. For this case, $K_L A_b = 0.063 \, \text{hr}^{-1}$ and $K_L A_s = 0.064 \, \text{hr}^{-1}$.

The air flow of the coarse bubble aeration system was then adjusted for the lower BOD of 30 mg/L that is assumed to occur at a depth of 73 m, with each of the 1,160 diffusers maintaining 2 mg/L in a 341 m^2 area. An air flow rate of 29 scmh per diffuser (33,640 scmh total) was determined to be sufficient, assuming that the characteristic equations may be extrapolated to 73 m of depth. For this case, $K_L A_b = 0.018 \, \text{hr}^{-1}$ and $K_L A_s = 0.003 \, \text{hr}^{-1}$. During the high BOD period at 10 m depth, approximately 3.5 times as much air flow is needed to maintain aerobic conditions.

A fine bubble aeration system for McCook Reservoir was designed using the same procedure. The system was first designed for the high BOD of 80 mg/L and a depth of 10 m. It was determined that a fine bubble diffuser with an air flow rate of 20 scmh would maintain aerobic conditions for an area of 115 m^2. From this, 3,438 fine bubble diffusers supplying a total air flow of 68,760 scmh would be needed for the reservoir. For this case, $K_L A_s = 0.003 \, \text{hr}^{-1}$ and $K_L A_s = 0.044 \, \text{hr}^{-1}$.

The air flow of the fine bubble system was then adjusted for the lower BOD of 30 mg/L and 73 m depth. With each of the 3,438 diffusers maintaining 2 mg/L in a 115 m^2 area, an air flow rate of 3.8 scmh per diffuser (13,065 scmh total) was determined. For this case, $K_L A_b = 0.034 \, \text{hr}^{-1}$ and $K_L A_s = 0.001 \, \text{hr}^{-1}$. During the high BOD period, approximately 5.3 times as much air flow is needed to maintain aerobic conditions.

A coarse bubble aeration system for McCook Reservoir requires only 1,160 diffusers, approximately one-third of the 3,438 diffusers needed for a fine bubble aeration system. However, significantly less air flow is needed for the fine bubble diffusers in comparison with the coarse bubble diffusers. At the more common depth of 10 m, the air flow required by the fine bubble diffusers was 39% of that required by the coarse bubble diffusers. These considerations and the mixing requirements of the reservoir are a part of the aeration system design.

Although the process to determine $K_L A_s$ and K_{Ls} is possible with a spread sheet, it is cumbersome for commercial specifications. The guideline for the testing of commercial aeration devices has been well developed and is generally available (American Society of Civil Engineers, 1992). There is no requirement to measure total dissolved gas pressure or estimate dissolved nitrogen concentration, and an

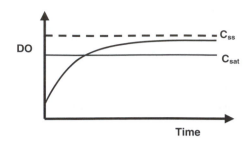

Figure 9.11. Disturbed equilibrium for the determination of the coefficients $K_L a$ and C_{ss}. DO is dissolved oxygen concentration, and C_{sat} is the saturation concentration at equilibrium with the atmosphere, which is generally below C_{ss}.

equation similar to equation (9.11) is fit to concentration versus time data for diffuser measurements in clean water:

$$\frac{\partial C}{\partial t} = -K_L a \left(C_{ss} - C \right) \tag{9.35}$$

with solution

$$\ln \left[\frac{C_{ss} - C}{C_{ss} - C(t = 0)} \right] = -K_L a t \tag{9.36}$$

where C is the dissolved oxygen concentration, $K_L a$ is a coefficient that represents liquid film coefficient times surface area divided by the volume of the water body, and C_{ss} is the steady-state concentration of dissolved oxygen that is determined from the fitting procedure. The tests are run such that dissolved oxygen concentration is reduced chemically, by adding sodium sulfite, or physically, such as through stripping with nitrogen gas, as illustrated in Figure 9.11. The dissolved oxygen concentration is recorded as it moves from the temporary reduction back to a steady-state concentration, C_{ss}. The values of $K_L a$ and C_{ss} are curve fit to these measurements. Either equation (9.35) or (9.36) can be fit to a concentration curve like the one given in Figure 9.11, varying C_{ss} and $K_L a$ until an optimum fit is achieved. The drawback to this technique is that $K_L a$ represents an unknown combination of $k_{LS} A_s/V$, $k_{Lb} A_b/V$, and depth, so the tests must be at the application depth.

Of course, both of the two coefficients, C_{ss} and $K_L a$ are some combination of the processes considered when equation (8.87) through (8.102) were developed, and are a function of liquid film coefficient across both the bubbles and the free surface, bubble and water surface interfacial area, hydrostatic pressure, the mole ratio of gas in the bubbles, and equilibrium with the atmosphere. These two coefficients, however, can be valuable in the design of an aeration system, as long as (1) the arrangement of diffusers in the water body or tank is similar to the application and (2) the depth of the test is the same as the application. Significant deviations from these two criteria will cause errors in the application of the tests to the field.

E. Problems

1. The effluent from a sewage treatment plant on a river with an upstream biochemical oxygen demand (BOD) of 2 g/m³ has a discharge of 10 m³/s and a BOD of 15 g/m³. Determine the best choice of stream reaeration coefficient,

and compute and plot the dissolved oxygen profile in the river downstream of the plant. Compute and plot the same with your justified high and low estimates of reaeration coefficient. Comment on the results.

Stream Characteristics:

$U = 0.5$ m/s
$Q = 100$ m³/s upstream of plant
Slope $= 10^{-4}$
Mean depth $= 2$ m
D_L is small
Width $= 220$ m
Water temperature $= 25\,°C$
Upstream dissolved oxygen concentration $= 6$ g/m³

2. A 10-km-long lake is exposed to various wind speeds. On a cold day in the fall, the water temperature at 10°C has not cooled yet, but the air temperature is at −5 °C. The relative humidity is 100%. Compute the gas film coefficient for water vapor at various wind fetch lengths for wind speeds at 2 m height and at 2, 6, 10, and 16 m/s. Compare your results with those of Example 8.6.

3. Data below were taken for an aerator test with air injected in an 8-m diameter tank of clean water at 30°C that was 9 m deep, with an aerator placed 8.5 m deep. The air flow rate is 51 standard cubic meters per hour.

   a. Estimate $K_L a$ and C_{ss}.

   b. Estimate $K_L A_b$ and K_{Ls}.

   c. Compare the two sets of estimates and discuss the differences.

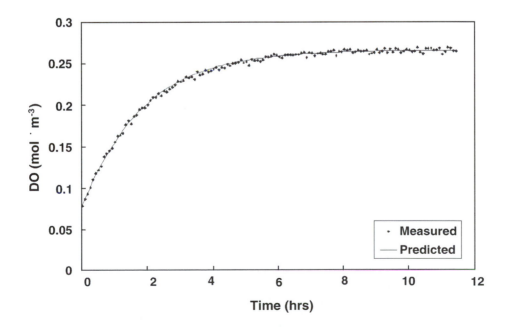

Moody's Diagram

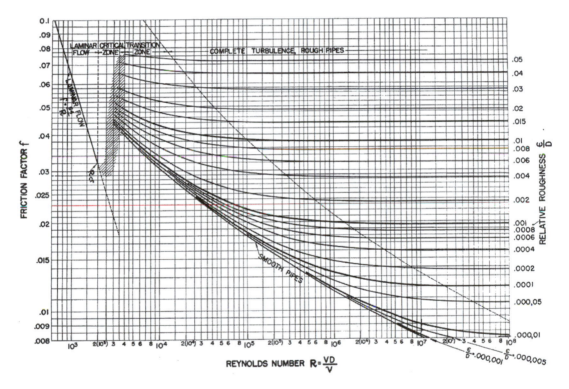

V is cross-sectional mean velocity, ν is kinematic viscosity, ε is wall roughness, and D is pipe diameter, or 4*Rh*. *Rh* is hydraulic radius, A/P, where A is cross-sectional area and P is wetted perimeter (Moody, 1944).

Scales of Pressure Measurement

Scale	Value at atmospheric pressure
Atmosphere	1
Pascal	101,325
Bar	1.013
Pounds/in^2	14.7
m H$_2$O ($\rho = 1000$ kg/m^3)	10.34
mm Hg ($\rho = 13,600$ kg/m^3)	760
ft H$_2$O	33.91
in Hg	29.92

Properties of Dry Air at Atmospheric Pressure

Temperature (°C)	Density (kg/m^3)	Kinematic viscosity (m^2/s)	Thermal diffusion coefficient (m^2/s)
0	1.292	1.33E-05	1.89E-05
5	1.269	1.37E-05	1.95E-05
10	1.246	1.42E-05	2.02E-05
15	1.225	1.46E-05	2.08E-05
20	1.204	1.51E-05	2.15E-05
25	1.183	1.55E-05	2.22E-05
30	1.164	1.60E-05	2.28E-05
35	1.146	1.64E-05	2.35E-05
40	1.128	1.69E-05	2.42E-05
45	1.11	1.74E-05	2.49E-05

APPENDIX A–4

Properties of Pure Water

Temperature (°C)	Density (kg/m³)	Kinematic viscosity (m²/s)	Vapor pressure (kPa)	Thermal diffusion coefficient (m²/s)
0	999.9	1.79E-6	0.61	1.34E-7
5	1000	1.52E-6	0.87	1.36E-7
10	999.7	1.31E-6	1.23	1.38E-7
15	999.1	1.14E-6	1.70	1.40E-7
20	998.2	1.01E-6	2.34	1.42E-7
25	997.1	8.97E-7	3.17	1.44E-7
30	995.7	8.04E-7	4.24	1.46E-7
35	994.1	7.27E-7	5.69	1.49E-7
40	992.2	6.61E-7	7.45	1.52E-7

The Error Function

x	erf x	x	erf x	x	erf x	x	erf x
0.00	0.0000	0.50	0.5205	1.00	0.8427	1.50	0.9661
0.01	0.0113	0.51	0.5292	1.01	0.8468	1.51	0.9673
0.02	0.0226	0.52	0.5379	1.02	0.8508	1.52	0.9684
0.03	0.0338	0.53	0.5465	1.03	0.8548	1.53	0.9695
0.04	0.0451	0.54	0.5549	1.04	0.8586	1.54	0.9706
0.05	0.0564	0.55	0.5633	1.05	0.8624	1.55	0.9716
0.06	0.0676	0.56	0.5716	1.06	0.8661	1.56	0.9726
0.07	0.0789	0.57	0.5798	1.07	0.8698	1.57	0.9736
0.08	0.0901	0.58	0.5879	1.08	0.8733	1.58	0.9745
0.09	0.1013	0.59	0.5959	1.09	0.8768	1.59	0.9755
0.10	0.1125	0.60	0.6039	1.10	0.8802	1.60	0.9763
0.11	0.1236	0.61	0.6117	1.11	0.8835	1.61	0.9772
0.12	0.1348	0.62	0.6194	1.12	0.8868	1.62	0.9780
0.13	0.1459	0.63	0.6270	1.13	0.8900	1.63	0.9788
0.14	0.1569	0.64	0.6346	1.14	0.8931	1.64	0.9796
0.15	0.1680	0.65	0.6420	1.15	0.8961	1.65	0.9804
0.16	0.1790	0.66	0.6494	1.16	0.8991	1.66	0.9811
0.17	0.1900	0.67	0.6566	1.17	0.9020	1.67	0.9818
0.18	0.2009	0.68	0.6638	1.18	0.9048	1.68	0.9825
0.19	0.2118	0.69	0.6708	1.19	0.9076	1.69	0.9832
0.20	0.2227	0.70	0.6778	1.20	0.9103	1.70	0.9838
0.21	0.2335	0.71	0.6847	1.21	0.9130	1.71	0.9844
0.22	0.2443	0.72	0.6914	1.22	0.9155	1.72	0.9850
0.23	0.2550	0.73	0.6981	1.23	0.9181	1.73	0.9856
0.24	0.2657	0.74	0.7047	1.24	0.9205	1.74	0.9861
0.25	0.2763	0.75	0.7112	1.25	0.9229	1.75	0.9867
0.26	0.2869	0.76	0.7175	1.26	0.9252	1.76	0.9872
0.27	0.2974	0.77	0.7238	1.27	0.9275	1.77	0.9877
0.28	0.3079	0.78	0.7300	1.28	0.9297	1.78	0.9882
0.29	0.3183	0.79	0.7361	1.29	0.9319	1.79	0.9886
0.30	0.3286	0.80	0.7421	1.30	0.9340	1.80	0.9891
0.31	0.3389	0.81	0.7480	1.31	0.9361	1.81	0.9895

(continued)

Appendix A–5 *(continued)*

x	**erf** x	x	**erf** x	x	**erf** x	x	**erf** x
0.32	0.3491	0.82	0.7538	1.32	0.9381	1.82	0.9899
0.33	0.3593	0.83	0.7595	1.33	0.9400	1.83	0.9903
0.34	0.3694	0.84	0.7651	1.34	0.9419	1.84	0.9907
0.35	0.3794	0.85	0.7707	1.35	0.9438	1.85	0.9911
0.36	0.3893	0.86	0.7761	1.36	0.9456	1.86	0.9915
0.37	0.3992	0.87	0.7814	1.37	0.9473	1.87	0.9918
0.38	0.4090	0.88	0.7867	1.38	0.9490	1.88	0.9922
0.39	0.4187	0.89	0.7918	1.39	0.9507	1.89	0.9925
0.40	0.4284	0.90	0.7969	1.40	0.9523	1.90	0.9928
0.41	0.4380	0.91	0.8019	1.41	0.9539	1.91	0.9931
0.42	0.4475	0.92	0.8068	1.42	0.9554	1.92	0.9934
0.43	0.4569	0.93	0.8116	1.43	0.9569	1.93	0.9937
0.44	0.4662	0.94	0.8163	1.44	0.9583	1.94	0.9939
0.45	0.4755	0.95	0.8209	1.45	0.9597	1.95	0.9942
0.46	0.4847	0.96	0.8254	1.46	0.9611	1.96	0.9944
0.47	0.4937	0.97	0.8299	1.47	0.9624	1.97	0.9947
0.48	0.5027	0.98	0.8342	1.48	0.9637	1.98	0.9949
0.49	0.5117	0.99	0.8385	1.49	0.9649	1.99	0.9951

Henry's Law Constants and Percent Resistance to Transfer in the Liquid Phase

Compound	Mol. formula	Temp.	Henry's law constant	Percent	Resistance	In Liquid	Phase	Ref.
		$°C$	atm* m^3/gmol	$K_G/K_L =$	20	50	150	
Acenaphthene	$C_{12}H_{10}$	25	2.28E−04		15.7	31.8	58.3	2
Acenapthene	$C_{12}H_{10}$	25	1.46E−04		10.7	23.0	47.2	4
Air		25	1.29E+00		99.9	100.0	100.0	3
Aldrin	$C_{12}H_8Cl_6$	25	1.44E−05		1.2	2.9	8.1	1
Ammonia	NH_3	25	7.82E−05		6.0	13.8	32.4	3
Anthracene	$C_{14}H_{10}$	25	1.65E−03		57.4	77.1	91.0	2
Argon	Ar	25	7.18E−01		99.8	99.9	100.0	3
Arochlor 1221		25	2.28E−04		15.7	31.8	58.3	9
Arochlor 1242		25	3.43E−04		21.9	41.2	67.8	9
Arochlor 1248		25	4.40E−04		26.5	47.4	73.0	9
Arochlor 1254		25	2.83E−04		18.8	36.7	63.5	9
Arochlor 1260		25	3.36E−04		21.6	40.7	67.3	9
Arochlor 1268		25	4.15E−04		25.3	45.9	71.8	9
Arsenic hydride	AsH_3	25	1.12E−01		98.9	99.6	99.9	3
Benzene	C_6H_6	25	5.49E−03		81.8	91.8	97.1	2
Benzene	C_6H_6	20	4.39E−03		78.5	90.1	96.5	5
Benzene	C_6H_6	25	5.55E−03		81.9	91.9	97.1	4
g-BHC		20	1.58E−06		0.1	0.3	1.0	10
Biphenyl	$C_{12}H_{10}$	25	6.36E−04		34.2	56.5	79.6	2
Biphenyl	$C_{12}H_{10}$	25	4.08E−04		25.0	45.5	71.5	4
1,3-Butadiene	C_4H_2	25	6.94E−02		98.3	99.3	99.8	3
n-Butane	C_4H_{10}	25	8.22E−01		99.9	99.9	100.0	3
n-Butane	C_4H_{10}	25	9.47E−01		99.9	99.9	100.0	2
1-Butanol	$C_4H_{10}O$	20	5.57E−06		0.5	1.1	3.4	5
1-Butene	C_4H_8	25	2.68E−01		99.5	99.8	99.9	2
1-Butyne	C_4H_6	25	1.94E−02		94.1	97.5	99.2	2
Carbon dioxide	CO_2	25	2.95E−01		99.6	99.8	99.9	3
Carbon monoxide	CO	25	1.05E+00		99.9	100.0	100.0	3
Carbon tetrachloride	CCl_4	20	2.27E−02		95.0	97.9	99.3	5
Carbon tetrafluoride	CF_4	25	4.73E+00		100.0	100.0	100.0	3

(continued)

Appendix A–6 *(continued)*

Compound	Mol. formula	Temp. °C	Henry's law constant atm* m³/gmol	Percent $K_G/K_L =$	Resistance In Liquid Phase 20	50	150	Ref.
Carbonyl sulfide	COS	25	4.71E−02		97.5	99.0	99.7	3
Chlordane	$C_{10}H_6Cl_8$	20	9.64E−05		7.4	16.7	37.6	10
Chlorine gas	Cl_2	25	1.09E−02		89.9	95.7	98.5	3
Chlorobenzene	C_6H_5Cl	20	2.61E−03		68.5	84.4	94.2	5
Chlorobenzene	C_6H_5Cl	25	3.77E−03		75.5	88.5	95.9	4
Chloroform	CCl_4	20	3.33E−03		73.5	87.4	95.4	6
Chloroform	CCl_4	20	2.42E−02		95.3	98.1	99.3	8
Chloroxide	ClO_2	25	9.72E−04		44.3	66.5	85.6	3
Cyclohexane	C_6H_{12}	25	1.96E−01		99.4	99.8	99.9	2
Cyclopentane	C_5H_{10}	25	1.87E−01		99.4	99.7	99.9	2
Cyclopropane	$c-C_3H_6$	25	8.70E−02		98.6	99.4	99.8	3
DBP		20	1.59E−06		0.1	0.3	1.0	10
p,p'-DDT		20	1.31E−05		1.1	2.7	7.6	10
Decane	$C_{10}H_{22}$	25	4.93E+00		100.0	100.0	100.0	2
DEHP		20	4.40E−06		0.4	0.9	2.7	10
1,2-Dibromoethane	$C_2H_4Br_2$	20	6.29E−04		34.4	56.7	79.7	5
1,2-Dibromoethane	$C_2H_4Br_2$	25	8.20E−04		40.1	62.6	83.4	11
Dichlorodifluoromethane	Cl_2F_2	20	1.49E+00		99.9	100.0	100.0	8
1,2-Dichloropropane	$C_3H_6Cl_2$	20	2.07E−03		63.3	81.2	92.8	5
Dieldrin	$C_{12}H_8Cl_6O$	20	7.80E−06		0.6	1.6	4.6	10
Difluorochloromethane	CHF_2Cl	25	2.89E−02		95.9	98.3	99.4	3
Difluorovinylidene	C_2F_2	25	6.33E−01		99.8	99.9	100.0	3
Dimethyl mercury	HgC_2H_6	25	7.70E−03		86.3	94.0	97.9	7
2,2-Dimethylpentane	C_7H_{16}	25	3.15E+00		100.0	100.0	100.0	2
Dodecane	$C_{12}H_{26}$	25	7.12E+00		100.0	100.0	100.0	2
Ethane	C_2H_6	25	5.40E−01		99.8	99.9	100.0	3
Ethane	C_2H_6	25	4.99E−01		99.8	99.9	100.0	2
Ethene	C_2H_4	25	2.10E−01		99.4	99.8	99.9	3
Ethene	C_2H_4	25	2.14E−01		99.4	99.8	99.9	2
Ethylbenzene	C_8H_{10}	25	8.73E−03		87.7	94.7	98.2	2
Ethylbenzene	C_8H_{10}	25	8.43E−03		87.3	94.5	98.1	4
Ethyne	C_2H_2	25	2.41E−02		95.2	98.0	99.3	3
9H-Fluorine	$C_{13}H_{10}$	25	2.35E−04		16.1	32.5	59.0	2
Helium	He	25	2.59E+00		100.0	100.0	100.0	3
n-Heptane	C_7H_{16}	25	2.07E+00		99.9	100.0	100.0	2
2-Heptanone	$C_7H_{14}O$	20	9.00E−05		7.0	15.8	36.0	5
2-Heptene	C_7H_{14}	25	4.18E−01		99.7	99.9	100.0	2
n-Hexane	C_6H_{14}	25	1.85E+00		99.9	100.0	100.0	2
1-Hexene	C_6H_{12}	25	4.12E−01		99.7	99.9	100.0	2
Hydrogen gas	H_2	25	1.28E+00		99.9	100.0	100.0	3
Hydrogen selenide	H_2Se	25	1.19E−02		90.7	96.1	98.6	3
Hydrogen sulfide	H_2S	25	9.74E−03		88.8	95.2	98.4	3
Isobutane	C_4H_{10}	25	1.23E+00		99.9	100.0	100.0	3
Isobutane	C_4H_{10}	25	1.24E+00		99.9	100.0	100.0	2
Isopentane	C_5H_{12}	25	1.36E+00		99.9	100.0	100.0	2
Isopropylbenzene	C_9H_{12}	25	1.45E−02		92.2	96.7	98.9	2
Isopropylbenzene	C_9H_{12}	25	1.46E−02		92.3	96.8	98.9	1

Compound	Mol. formula	Temp.	Henry's law constant	Percent	Resistance	In Liquid	Phase	Ref.
		°C	atm* m³/gmol	$K_G/K_L =$	20	50	150	
Krypton	Kr	25	4.02E−01		99.7	99.9	100.0	3
Lindane		25	4.93E−07		0.0	0.1	0.3	1
Mecuric cloride	$HgCl_2$	25	6.70E−10		0.0	0.0	0.0	7
Mecury	Hg	25	7.80E−03		86.5	94.1	98.0	7
Mercury	Hg	25	1.14E−02		90.3	95.9	98.6	1
Methane	CH_4	25	7.20E−01		99.8	99.9	100.0	3
Methane	CH_4	25	6.65E−01		99.8	99.9	100.0	2
Methyl bromide	CH_3Br	25	6.15E−03		83.4	92.6	97.4	3
Methyl chloride	CH_3Cl	20	2.75E−03		69.6	85.1	94.5	6
Methyl chloride	CH_3Cl	25	9.60E−03		88.7	95.2	98.3	3
Methyl chloroform	$C_4H_3Cl_3$	20	1.49E−02		92.5	96.9	98.9	8
Methyl floride	CH_3F	25	1.70E−02		93.3	97.2	99.1	3
Methyl mercuric cloride	$HgClCH_3$	25	5.00E−07		0.0	0.1	0.3	7
2-Methyl-1-propanol	$C_4H_{10}O$	20	1.03E−05		0.8	2.1	6.0	5
Methylcyclohexane	C_7H_{16}	25	4.28E−01		99.7	99.9	100.0	2
Methylcyclopentane	C_6H_{12}	25	3.62E−01		99.7	99.9	100.0	2
3-Methylheptane	C_8H_{18}	25	3.71E+00		100.0	100.0	100.0	2
2-Methylhexane	C_7H_{16}	25	3.42E+00		100.0	100.0	100.0	2
4-Methyloctane	C_9H_{20}	25	9.94E+00		100.0	100.0	100.0	2
2-Methylpentane	C_6H_{14}	25	1.73E+00		99.9	100.0	100.0	2
2-Methylpropene	C_4H_8	25	1.75E−01		99.3	99.7	99.9	3
Napthalene	$C_{10}H_8$	25	4.25E−04		25.8	46.5	72.3	2
Napthaline	$C_{10}H_8$	25	4.83E−04		28.3	49.7	74.8	4
Neon	Ne	25	2.22E+00		99.9	100.0	100.0	3
Neopentane	C_5H_{10}	25	1.68E+00		99.9	100.0	100.0	3
Nitric oxide	NO	25	5.19E−01		99.8	99.9	100.0	3
Nitrogen gas	N_2	25	1.54E+00		99.9	100.0	100.0	3
Nitrogen trifluoride	NF_3	25	1.26E+00		99.9	100.0	100.0	3
Nitrous oxide	N_2O	25	4.12E−02		97.1	98.8	99.6	3
n-Nonane	C_9H_{20}	25	3.29E+00		100.0	100.0	100.0	2
n-Octane	C_8H_{18}	25	3.22E+00		100.0	100.0	100.0	2
1-Octene	C_8H_{16}	25	9.05E−01		99.9	99.9	100.0	2
Oxygen gas	O_2	25	7.86E−01		99.8	99.9	100.0	3
n-Pentane	C_5H_{12}	25	1.26E+00		99.9	100.0	100.0	2
1-Pentanol	$C_5H_{12}O$	20	1.03E−05		0.8	2.1	6.0	5
2-Pentanone	$C_5H_{10}O$	20	3.16E−05		2.6	6.2	16.5	5
1-Pentene	C_5H_{10}	25	3.98E−01		99.7	99.9	100.0	2
Perafluorohydrazine	C_3F_6	25	3.41E+00		100.0	100.0	100.0	3
Perchloroethene	C_2Cl_4	20	1.51E−02		92.6	96.9	98.9	8
Phenanthrene	$C_{14}H_{10}$	25	1.48E−04		10.8	23.2	47.6	2
Phenanthrene	$C_{14}H_{10}$	25	3.93E−05		3.1	7.4	19.4	4
Propane	$n-C_3H_8$	25	6.68E−01		99.8	99.9	100.0	3
Propane	C_3H_8	25	7.07E−01		99.8	99.9	100.0	2
Propene	C_3H_6	25	1.35E−01		99.1	99.6	99.9	3
Propene	C_3H_6	25	2.32E−01		99.5	99.8	99.9	2
Propylcyclopentane	C_8H_{16}	25	8.93E−01		99.9	99.9	100.0	2

(continued)

Appendix A–6 *(continued)*

Compound	Mol. formula	Temp.	Henry's law constant	Percent	Resistance	In Liquid	Phase	Ref.
		°C	atm* m³/gmol	$K_G/K_L =$ 20	50	150		
Propyne	C_3H_4	25	1.46E−02	92.3	96.8	98.9	3
Propyne	C_3H_4	25	1.10E−02	90.0	95.7	98.5	2
Radon	Rn	25	1.08E−01	98.9	99.5	99.8	3
Sulfur dioxide	SO_2	25	5.56E−01	99.8	99.9	100.0	3
Sulfur hexafloride	SF_6	25	4.11E+00	100.0	100.0	100.0	3
Tetrachloroethene	C_2Cl_4	20	1.30E−02	91.5	96.4	98.8	6
Tetradecane	$C_{14}H_{30}$	25	1.14E+00	99.9	100.0	100.0	2
Tetrafluorohydrazine	N_2F_4	25	1.17E+00	99.9	100.0	100.0	3
Toluene	C_7H_8	25	6.66E−03	84.5	93.2	97.6	2
Toluene	C_7H_8	20	5.18E−03	81.2	91.5	97.0	5
Toluene	C_7H_8	25	6.64E−03	84.5	93.1	97.6	4
Toxaphene	$C_{10}H_{10}Cl_8$	20	2.38E−07	0.0	0.0	0.1	10
1,1,1-Trichloroethane	C_2HCl_3	20	1.32E−02	91.7	96.5	98.8	6
Trichloroethene	$C_2H_3Cl_3$	20	7.64E−03	86.4	94.1	97.9	6
Trichloromethane	C_4HCl_3	20	5.39E−03	81.8	91.8	97.1	8
Trichloromethane	C_2HCl_3	20	9.86E−03	89.1	95.4	98.4	8
2,2,4-Trimethylpentane	C_8H_{18}	25	3.04E+00	100.0	100.0	100.0	2
Xenon	Xe	25	2.33E−01	99.5	99.8	99.9	3
o-Xylene	C_8H_{10}	25	5.27E−03	81.2	91.5	97.0	2

References:
1. Mackay and Leinonen (1975).
2. Mackay and Shiu (1981).
3. Wilhelm, Battino and Wilcock (1977).
4. Mackay, Shiu and Sutherland (1979).
5. Mackay and Yuen (1983).
6. Lincoff and Gosset (1984).
7. Mackay and Shiu (1984).
8. Munz and Roberts (1984).
9. Burkhard, et al. (1985).
10. Mackay, Peterson and Schroeder (1986).
11. Rathbun and Tai (1986).

References

Abernathy, R. B., Benedict, R. P., Dowdell, R. B. (1985). *J. Fluids Eng.*, 107, 161.

Adams, E. E., Cosler, D. J., and Helfrich, K. R. (1990). *Water Resources Res.*, 26(3), 425.

American Society of Civil Engineering (1992). *Standard for the Measurement of Oxygen Transfer in Clean Water*, ASCE, New York.

Asher, W. E., Karle, L. M., Higgins, B. J., Farley, P. J., and Monahan, E. C. (1996). *J. Geophys. Res.*, 101, 12027.

Asher, W. E., and Wanninkhof, R. (1998). *J. Geophys. Res.*, 103, 15993.

Atmane, M. A., Asher, W. E., and Jessup, A. T. (2004). *J. Geophys. Res.*, 109, C08S14.

Bird, R. B., Stewart, W. E., and Lightfoot, E. N. (1960). *Transport Phenomena*. John Wiley and Sons, New York.

Blasius, H. (1908). *Z. Angew. Math. Phys.*, 56, 1 (Engl. Trans NACA Tech. Mem. 1256).

Boussinesq, J. (1877). *Mem. Pres. Acad. Sci.*, Paris, 23, 46.

Bowen, I. S. (1926). *Phys. Rev.*, 27, 1.

Brady, K. D., Graves, W. L. and Geyer, J. C. (1969). *Surface Heat Exchange at Power Plant Cooling Lakes*. Report RP-49, Johns Hopkins University, Baltimore, MD.

Broecker, W. S., and Peng, T. H. (1982). *Tracers in the Sea*. Eldigio, Palisades, NY.

Brumley, B. B., and Jirka, G. H. (1987). *J. Fluid Mech.*, 183, 235.

Brutsaert, W. H. (1982). *Evaporation into the Atmosphere – Theory, History and Application*. D. Reidel, Hingham, MA.

Calderbank, P. H. (1967). In *Mixing*, vol. 2. Edited by U. Uhl and J. B. Gray. Academic Press, San Diego, p. 1.

Campbell, J. A., and Hanratty, T. J. (1982). *J. Am. Inst. Chem. Eng.*, 28, 988.

Chan, W. E., and Scriven, L. E. (1970). *Ind. and Eng. Chem. Fund.*, 9, 114.

Chapman, S., and Cowling, T. G. (1970). *The Mathematical Theory of Non-Uniform Gases*, 3rd ed. Cambridge University Press, Cambridge.

Clarke, J. F., Wanninkhof, R., Schlosser, P., and Simpson, H. J. (1994). *Tellus, Ser. B*, 46, 274.

CRC (2005). *Handbook of Chemistry and Physics*. CRC Press, Boca Raton, FL.

Cussler, E. L. (1976). *Multicomponent Diffusion*. Elsevier, New York.

Cussler, E. L. (1997). *Diffusion: Mass Transfer in Fluid Systems*, 2nd ed. Cambridge University Press, Cambridge.

Danckwerts, P. V. (1951). *Ind. Eng. Chem.*, 23(6), 1460.

Danckwerts, P. V. (1970). *Gas-Liquid Reactions*. McGraw-Hill, New York.

Daniil, E. I., and Gulliver, J. S. (1991). *J. Environ. Eng.*, 117, 522.

Davies, J. T., and Khan, W. (1965). *Chem. Eng. Sci.*, 20, 713.

Duffy, P. B., and Caldeira, K. (1995). *Global Biogeochem. Cycles*, 9, 373.

Einstein, A. (1905). *Annalel der Physik*, 17, 549.

Elder, J. W. (1959). *J. Fluid Mech.*, 5, 544.

Fick, A. E. (1855). *Poggendorff's Annelen der Physic*, 94, 59.

Ficke, J. F. (1972). *U.S. Geol. Surv. Prof. Pap.*, 65.

Fisher, H. B. (1967a). *J. Hydraul. Div. Am. Soc. Civ. Eng.*, 93, 187.

Fisher, H. B. (1967b). *U.S. Geol. Surv. Prof. Pap. 575-D*.

Fisher, H. B. (1973). *Ann. Rev. Fluid Mech.*, 5, 59.

Fisher, H. B., List, J. E., Koh, R. C. Y., Imberger, J., and Brooks, N. H. (1979). *Mixing in Inland and Coastal Waters*. Academic Press, San Diego.

Freeze, R. A., and Cherry, J. A. (1979). *Groundwater*. Prentice Hall, Englewood Cliffs, NJ.

Frew, N. M. (1997). In *The Sea Surface and Global Change*. Edited by R. A. Duce and P. S. Liss. Cambridge University Press, New York, 121.

Friedlander, S. K. (1961). *AIChE J.*, 7, 317.

Gelhar, L., Welty, C., and Rehfeldy, K. R. (1992). *Water Res. Res.*, 28(7), 1955.

Glover, R. E. (1964). *U.S. Geol. Surv. Prof. Pap. 433-B*.

Godfrey, R. G., and Frederick, B. J. (1970). *U.S. Geol. Surv. Prof. Pap. 433-K*.

Gulliver, J. S. (1977). *Analysis of Surface Heat Exchange and Longitudinal Dispersion in a Narrow Open Field Channel with Application to Water Temperature Prediction*, M.S. thesis, University of Minnesota.

Gulliver, J. S., and Halverson, M. J. (1989). *Water Res. Res.*, 25, 1783.

Gulliver, J. S., and Song, C. S. S. (1986). *J. Geophys. Res.*, 91(C4), 5145.

Gulliver, J. S., and Stefan, H. G. (1986). *J. Environ. Eng. Div., Am. Soc. Civ. Eng.*, 112(2), 387.

Gulliver, J. S., Erickson, B., Zaske, A. J., and Shimon, K. S. (2002). In *Gas Transfer at Water Surfaces*. Edited by M. Donelan, W. Drennan, E. Monehan, and R. Wanninkhof. Geophysical Monograph 127, American Geophysical Union, Washington, DC.

Harbeck, G. E., Jr. (1962). *U.S. Geol. Surv., Prof. Pap. 272E*.

Harbeck G. E., Jr., Kohler, M. A., and Koberg, G. E. (1958). *U.S. Geol. Surv., Prof. Pap. 298*.

Haußecker, H., Reinelt, S., and Jähne, B. (1995). In *Air-Water Gas Transfer*. Edited by B. Jähne and E. C. Monahan. Aeon Verlag, Hanau, Germany, 405.

Hayduk, W., and Laudie, H. (1974). *Am. Inst. Chem. Eng. J.*, 20, 611.

Hibbs, D. E., Parkhill, K. L., and Gulliver, J. S. (1998). *J. Envir. Eng.*, 124(8), 752.

Higbie, R. (1935). *Am. Inst. Chem. Eng. J.*, 31, 365.

Hirschfelder, J., Curtiss, C. F., and Bird, R. B. (1954). *Molecular Theory of Gases and Liquids*. John Wiley and Sons, New York.

Hornix, W. J., and Mannaerts, S. H. W. M. (2001). *Van't Hoff and the Emergence of Chemical Therodynamics*. Coronet Books, Inc., Philadelphia.

Hsu, S. A. (1974). *J. Phys. Oceanogr.*, 4, 116.

Hughes, G. H. (1967). *U.S. Geol. Surv., Prof. Pap. 272H*.

Jähne, B. (1991). In *Air-Water Mass Transfer*. Edited by S. C. Wilhelms and J. S. Gulliver. American Society of Civil Engineers, Reston, VA, 582.

Jähne, B., Fischer, K. H., Imberger, J., Libner, P., Weiss, W., Imboden, D., Lemnin, U., and Jaquet, J. M. (1984). In *Gas Transfer at Water Surfaces*. Edited by W. Brutsaert and G. H. Jirka. D. Reidel, Norwell, MA, 303.

Jähne, B., Munich, K. O., Bosinger, R., Dutzi, A., Huber, W., and Libner, P. (1987). *J. Geophys. Res.*, 91, 1937.

Karikhoff, S. W., Brown, D. S., and Scott, T. A. (1979). *Water Res.*, 13, 241.

Kauzmann, W. (1966). *Kinetic Theory of Gases*. W.A. Benjamin, New York.

Kilpatrick, F. A., Rathbun, R. E., Yotsukura, N., Parker, G. W., and DeLong, L. L. (1979). *Techniques of Water Resources Investigations of the U.S. Geological Survey*, Book 3, Chapter A18. U.S. Geological Survey, Washington, DC.

Koch, D. L. and Brady, J. F. (1985). *J. Fluid Mech.*, 154, 399.

Kohler, M. A. (1954). *U.S. Geol. Surv., Prof. Pap. 169*.

Komori, S., Ueda, H., Ogino, F., and Mizushina, T. (1982). *Int. J. Heat Mass Transfer*, 25(4), 513.

Krenkel, P. A., and Orlob, G. T. (1962). *J. Sanit. Eng. Div. Am. Soc. Civ. Eng.*, 88(SA2), 53.

Kreyszig, E. (1982). *Advanced Engineering Mathematics*, 4th ed. John Wiley and Sons, New York.

Lau, Y. L. (1975). *Prog. Water Technol.*, 7(3/4), 519.

Law, C. N. S., and Khoo, B. C. (2002). *AIChE J.*, 48(9), 1856.

LeBas, G. (1915). *The Molecular Volume of Liquid Chemical Compounds*. Longmans, Green, NY.

Leighton, D. T., and Calo, J. M. (1981). *J. Chem. Eng. Data*, 6, 87.

Levenspiel, O. (1962). *Chemical Reaction Engineering*. John Wiley and Sons, New York.

Levich, V. G. (1962). *Physicochemical Hydrodynamics*. Prentice Hall, Englewood Cliffs, NJ.

Liss, P. S., and Merlivat, L. (1986). In *The Role of Air-Sea Exchange in Geochemical Cycling*. Edited by P. Buat-Menard. D. Reidel, Norwell, MA, 113.

Livingstone, D. M., and Imboden, D. M. (1993). *Tellus, Ser. B*, 45, 275.

Long, S. R., and Hwang, N. E. (1976). *J. Fluid Mech.*, 77, 209.

Lyman, W. J., Reehl, W. F., and Rosenblatt, D. H. (1990). *Handbook of Chemical Property Estimation*. American Chemical Society, Washington, DC.

Mackay, D., and Yuen, A. T. K. (1983). *Environ. Sci. Tech.*, 17, 211.

Marciano, J. J., and Harbeck, G. E. (1954). *U.S. Geol. Surv. Prof. Pap. 267*.

McCabe, W. L., and Smith, J. C. (1975). *Unit Operations of Chemical Engineering*, 3rd ed. McGraw-Hill, New York.

McCready, M. A., Vassiliadou, E., and Hanratty, T. J. (1986). *AIChE J.* 32(7), 1108.

McKenna, S. P., and McGillis, W. R. (2004). *Int. J. Heat Mass Transfer*, 47(3), 539.

McQuivey, R. S., and Keefer, T. N. (1974). *J. Env. Eng. Div., Am. Soc. Civ. Eng.*, 100(4), 997.

McWhirter, J. R., and Hutter, J. C. (1989). *AIChE J.* 35(9), 1527.

Mitsuyatsu, H. (1968). *Rep. Res. Inst. Appl. Mech., Kyushu Univ.*, 16(55), 459.

Moody, L. F. (1944). *Trans. Am. Soc. Mech. Eng.*, 66, 671.

Moog, D. B., and Jirka, G. H. (1998). *J. Environ. Eng.*, 124(2), 104.

Munz, C., and Roberts, P. V. (1984). In *Gas Transfer at Water Surfaces*. Edited by W. Brutzert and G. H. Jirka. D. Reidel, Dordrecht, 35.

Nernst, W. (1904). *Zeitschrift für Physikalische Chemie*, 47, 52.

Nezu, I., and Nakagawa, H. (1993). *Turbulence in Open Channel Flow*. Balkema, Rotterdam, The Netherlands.

Nirmalakhandan, N., Brennan, R. A., and Speece, R. E. (1997). *Wat. Res.*, 31, 1471.

Orlins, J. J., and Gulliver, J. S. (2002). In *Gas Transfer at Water Surfaces*, Geop. Mono. 127, 247. Edited by M. A. Donelan, W. M. Drennan, E. S. Salzman, and R. Wanninkhof. American Geophysical Union, Washington, DC.

Owens, M., Edwards, R. W., and Gibbs, J. W. (1964). *Int. J. Air Water Pollut.*, 8, 469.

Patankar, S. V. (1980). *Numerical Heat Transfer and Fluid Flow*. Hemisphere, New York.

Prandtl, L. (1925). *Z. Angew. Math Mech.*, 5, 136–139.

Prandtl, L. (1942). *Z. Angew. Math. Mech.*, 22, 241–243.

Rashidi, M., and Banerjee, S. (1988). *Phys. Fluids*, 31(9), 2491.

Reichardt, H. (1951). *Ann. Angew. Math. Mech.*, 31, 7.

Reid, R. C., Sherwood, T. K., and Prausnitz, J. M. (1977). *Properties of Gases and Liquids*, 3rd ed. McGraw-Hill, New York.

Reynolds O. (1895). *Phil. Trans. R. Soc., Ser. A*, 186, 123–264.

Robertson, D. M. (2000). *U.S. Geol. Surv. Water-Resources Investigations Report 00-4258*, Washington, DC.

Rouse, H. (1937). *Transactions*, 102, 463.

Ryan, P. J., and Harleman, D. R. F. (1973). *T.R. 161*, R. M. Parsons Laboratory, Massachusetts Institute of Technology, Cambridge, MA.

Rymsha, V. A., and Dochenko, R. V. (1958). *Turdy GG1*, 65, 1.

Schierholz, E. L., Gulliver, J. S., Wilhelms, S. C., and Henneman, H. E. (2006). *Water Res.*, 40,1018.

Schlicting, H. (1979). *Boundary Layer Theory*, 7th ed. McGraw-Hill, New York.

Schuster, J. C. (1965). *J. Hydraul. Div. Am. Soc. Civ. Eng.*, 91, 101.

Sherwood, T. K., Pigford, R. L., and Wilke, C. R. (1975). *Mass Transfer*. McGraw-Hill, New York.

Shulyakovskyi, L. G. (1969). *Sov. Hydrol. Selec. Pap.*, 6, 566.

Sikar, T. T., and Hanratty, T. J. (1970). *J. Fluid Mech.*, 44, 589.

Spaulding, D. B. (1972). *Int. J. Numerial Methods in Eng.*, 4, 551.

State of California (1962). *Sacramento River Water Pollution Survey Bill No. 111*, Department of Water Resources, Sacramento.

Stefan, H. G., and Gulliver, J. S. (1978). *J. Environ. Eng. Div., Proc. Am. Soc. Civ. Eng.*, 104(EE2), 199.

Stokes, G. G. (1851). *Trans. Camb. Phil. Soc.*, 9(Part II), 8.

Streeter, H. W., and Phelps, E. B. (1925). *Publ. Health Bulletin No. 146*, U.S. Public Health Service.

Tamburrino, A., and Aravena, C. (2000). *XX Latin American Congress on Hydraulics*. International Association for Hydraulic Research, Rio de Janero.

Tamburrino, A., and Gulliver, J. S. (1994). In *Free-Surface Turbulence*. Edited by E. P. Rood and J. Katz. American Society of Mechanical Engineers, New York, FED-181.

Tamburrino, A., and Gulliver, J. S. (2002). *AIChE J.*, 48(12), 2732.

Taylor, G. I. (1953). *Proc. R. Soc. Lond., Ser. A*, 219, 186.

Taylor, G. I. (1954). *Proc. R. Soc. Lond., Ser. A*, 223, 186.

Thackston, E. L., and Krenkel, P. A. (1969). *J. Sanit. Eng. Div. Am. Soc. Civ. Eng.*, 95(SA1), 65.

Thomas, I. E. (1958). *Dispersion in Open-Channel Flow*, Ph.D. thesis, Northwestern University.

Treybal, R. E. (1980). *Mass Transfer Operations*, 3rd ed. McGraw-Hill, New York.

Tsivoglou, E. C., and Wallace, J. R. (1972). *Characterization of Stream Reaeration Capacity*, Report EPA-R3-72-012, U.S. Environmental Protection Agency, Washington, DC.

Torgersen, T. G., Mathieu, G., Hesslein, W., and Broecker, W. S. (1982). *J. Geophys. Res.*, 87, 546.

Turner, D. B. (1994). *Workbook of Atmospheric Dispersion Estimates*, 2nd ed. Lewis, Boca Raton, FL.

Turner, J. F. (1966). *U.S. Geol. Surv. Prof. Pap. 272-G*.

Upstill-Goddard, R. C., Watson, A. J., Liss, P. S., and Liddicoat, M. I. (1990). *Tellus, Ser. B*, 42, 364.

Veith, G. D., DeFoe, D. L., and Bergstedt, B. J. (1979). *J. Fish. Res. Board Can.*, 36, 1040.

Von Kármán, T. (1930). *Nachr. Ges. Wiss. Goett. Matn.-Phys.*, Kl, 58–76.

Wilke, C. R., and Lee, C. N. (1955). *Ind. Eng. Chem.*, 47, 1253.

Wanninkhof, R. (1992). *J. Geophys. Res.*, 97, 7373.

Wanninkhof, R., Asher, W. E., Weppernig, R., Chen, H., Schlosser, P., Langdon, C., and Sambrotto, R. (1993). *J. Geophys. Res.*, 98, 20237.

Wanninkhof, R., Ledwell, J. R., and Broecker, W. S. (1985). *Science*, 227, 1224.

Wanninkhof, R., Ledwell, J. R., Broecker, W. S., and Hamilton, M. (1987). *J. Geophys. Res.*, 92, 14, 567.

Watson, A. J., Upstill-Goddard, R. C., and Liss, P. S. (1991). *Nature*, 349, 145.

Wilke, C. R., and Chang, P. C. (1955). *J. Am. Inst. Chem. Eng.*, 2, 164.

Wu, J. (1975). *J. Fluid Mech.*, 68, 49.

Wu, J. (1980). *J. Geophys. Res.*, 74, 444.

Yotsukura, N., Fisher, H. B., and Sayre, W. W. (1970). *U.S. Geol. Surv. Water-Supply Pap. 1899-G*.

Subject Index

Index to Example Solutions